T0205631

Sustainable Textiles: Production, Processing, Manufacturing & Chemistry

Series Editor

Subramanian Senthilkannan Muthu, Head of Sustainability, SgT and API, Kowloon, Hong Kong

This series aims to address all issues related to sustainability through the lifecycles of textiles from manufacturing to consumer behavior through sustainable disposal. Potential topics include but are not limited to: Environmental Footprints of Textile manufacturing; Environmental Life Cycle Assessment of Textile production; Environmental impact models of Textiles and Clothing Supply Chain; Clothing Supply Chain Sustainability; Carbon, energy and water footprints of textile products and in the clothing manufacturing chain; Functional life and reusability of textile products; Biodegradable textile products and the assessment of biodegradability; Waste management in textile industry; Pollution abatement in textile sector; Recycled textile materials and the evaluation of recycling; Consumer behavior in Sustainable Textiles; Eco-design in Clothing & Apparels; Sustainable polymers & fibers in Textiles; Sustainable waste water treatments in Textile manufacturing; Sustainable Textile Chemicals in Textile manufacturing. Innovative fibres, processes, methods and technologies for Sustainable textiles; Development of sustainable, eco-friendly textile products and processes; Environmental standards for textile industry; Modelling of environmental impacts of textile products; Green Chemistry, clean technology and their applications to textiles and clothing sector; Eco-production of Apparels, Energy and Water Efficient textiles. Sustainable Smart textiles & polymers, Sustainable Nano fibers and Textiles; Sustainable Innovations in Textile Chemistry & Manufacturing; Circular Economy, Advances in Sustainable Textiles Manufacturing; Sustainable Luxury & Craftsmanship; Zero Waste Textiles.

More information about this series at https://link.springer.com/bookseries/16490

Subramanian Senthilkannan Muthu
Editor

Sustainable Approaches in Textiles and Fashion

Consumerism, Global Textiles and Supply Chain

 Springer

Editor
Subramanian Senthilkannan Muthu
Head of Sustainability
SgT and API
Kowloon, Hong Kong

ISSN 2662-7108   ISSN 2662-7116 (electronic)
Sustainable Textiles: Production, Processing, Manufacturing & Chemistry
ISBN 978-981-19-0876-7   ISBN 978-981-19-0874-3 (eBook)
https://doi.org/10.1007/978-981-19-0874-3

This Springer imprint is published by the registered company Springer Nature Singapore Pte Ltd.
The registered company address is: 152 Beach Road, #21-01/04 Gateway East, Singapore 189721,
Singapore

Contents

About the Editor

Dr. Subramanian Senthilkannan Muthu currently works for SgT Group as Head of Sustainability and is based out of Hong Kong. He earned his Ph.D. from The Hong Kong Polytechnic University and is a renowned expert in the areas of Environmental Sustainability in Textiles & Clothing Supply Chain, Product Life Cycle Assessment (LCA) and Product Carbon Footprint Assessment (PCF) in various industrial sectors. He has five years of industrial experience in textile manufacturing, research and development and textile testing and over a decade's experience in life cycle assessment (LCA), carbon and ecological footprints assessment of various consumer products. He has published more than 100 research publications, written numerous book chapters and authored/edited over 100 books in the areas of Carbon Footprint, Recycling, Environmental Assessment and Environmental Sustainability.

Global Textiles and Its Alignment with Sustainability

Mukta Ramchandani, Ivan Coste-Manière, Ishika Walia, Jun Wang, and Siqian Yang

Abstract Textile industry is one of the most traditional business sectors in our world, the social and environmental impacts of this profitable and important industry have long been an issue of global concern. With the world facing an unexpected crisis, caused by the Covid-19 outbreak, severe problems in the textile industry are unfolding dramatically. The whole industry is at the urge of transformation. In this chapter, we assess the status of the global textiles with focus on three regions (EU, China, and India), the challenges and opportunities emerging out of the pandemic, and its impact on the sustainable luxury fashion industry. Several industry examples are also discussed with future strategies to align with sustainability.

Keywords Sustainable fashion · Sustainable luxury · Fashion textiles · Textile sustainability · Textile manufacturing · Pandemic fashion · Sustainable development goals · Circular fashion

1 Introduction

Textile industry is not a single word, it involves actors from agricultural, chemical fibers, dyes, and chemical manufacturing, textile and apparel industry, retail and service sector, as well as waste treatment. The requirement of textile production dominates the huge amount globally since it represents the second essential requirement of humans. Meanwhile, the global textile industry performs as one of the largest polluters in this world by taking a disastrous toll on the environment in terms of fuel and chemical load to water resources, pesticide use in cotton farming, waste generation at the end-of-life stage, and more. Therefore, sustainable textiles mean that all materials and processes, inputs, and outputs, are healthy and safe for humans and the environment, in all phases of the product life cycle and all the energy, material, and process inputs come from renewable or recycled sources.

M. Ramchandani · I. Coste-Manière (✉) · I. Walia · J. Wang · S. Yang
SKEMA Business School, Luxury and Fashion Management, Université Côte d'Azur, 60 Rue Dostoievski, 6902 Sophia Antipolis Cedex, France
e-mail: ivan.costemaniere@skema.edu

S. S. Muthu (ed.), *Sustainable Approaches in Textiles and Fashion*,
Sustainable Textiles: Production, Processing, Manufacturing & Chemistry,
https://doi.org/10.1007/978-981-19-0874-3_1

In a recent report by [24], fossil fuel consumption by fashion companies needs to be reduced drastically as it poses an immense threat to the environment. Taking examples of the countries like Vietnam, China, and Turkey, the report highlights the issue of environmental degradation using coal in these countries for fashion textile productions. Further, the report illustrates how the energy needs for apparel production can be transformed through renewable sources of energy for a safe liveable future.

According to WTO statistics for 2020, China, the European Union, and India dominate the top three textile export regions. In this chapter, we investigate: how these three regions (China, India, and EU) implement sustainable development goals in the textile industry and how it impacts the global fashion industry? The current chapter also aims to showcase the need for sustainability in the global textiles sector and how decarbonizing is the need of the hour.

2 Methodology

In this chapter, the authors researched a variety of secondary sources and cases to delve into the topic of global textiles and sustainability. Latest business practices including examples from the pandemic are also discussed.

3 Environmental Sustainability

The sourcing of sustainable raw material is a crucial step for textile manufacturers. These raw materials must include eco-friendly production processes from yarn to fabric manufacturing, fabric chemical processing, effluent treatment of waste to name a few.

The current sustainable textile raw materials can be divided into four categories:

Plant-Based Fibers: The plant-based fibers produce lower environmental impact due to their very nature of biodegradability. While there can also be environmental concerns related to the amount of water used in producing natural fibers as seen in the case of organic cotton. There may be other natural alternatives that can be considered like hemp, bamboo fabric, lotus silk, banana fabric, etc.

Animal-Based Fibers: produced in a sustainable way. Some examples include vegetable-tanned leather, sustainable silk production, and sustainable wool.

Semi-Synthetic: with lower environmental impact and durability. Some examples include cotton and polyester or cotton and lycra, acrylic and wool, viscose, and spandex [10].

Recycled Fibers: These are made with waste material like recycled polyester [4], which may be frowned upon by sustainability experts. But from practicality use for outdoor apparel wear it still outperforms natural fabrics like cotton. Another example can be seen with Freitag's use of recycled truck tarps in the manufacturing of bags and accessories [9].

Economic sustainability in textile ensures that the business is achieving its targeted profitability, simultaneously using the resources in a sustainable manner (i.e., the business is not creating environmental concerns or using excessive resources). There is a steady increase in the number of firms considering their relationship to the community, to improve economic sustainability. Manufacturing of apparel and textile products in a country to be sold locally or globally has a direct or indirect influence on that country's economy. The sustainability practices in the apparel and textile sector should not ignore the dilemma of resource depletion. For the manufacturers to be more sustainable, life cycle assessments of the raw materials used and waste minimization should be of prime focus along with the biodegradability, repairability, and longevity aspects of the finished product [17].

4   Main Capacity and Consequence

When it comes to sustainability in the global textile arena, the global supply chain network is often neglected. The larger brands try to produce and source raw materials from first-tier suppliers that comply with the environmental norms and provide evidence of safe working conditions for the workers. However, these primary suppliers may work with contractual suppliers which account for the lower-tier suppliers. Often the low-tier suppliers do not comply with the environmental or safety norms, which may be since the environmental or sustainability regulations are non-existent, not enforced, or regulated in their country [26]. Another issue arising here is the consequences for smaller businesses. Not all companies can source materials globally from first-tier suppliers to be sure of sustainable practices. A solution can be to source materials from the low-tier suppliers by performing company audits and frequent visits to the suppliers. Both large and small businesses can collectively lead the way for the education and training of low-tier suppliers to direct them toward sustainable practices in those countries.

5   Challenges and Opportunities

5.1   Opportunities

Support from governing organization
Several international government organizations, non-governmental organizations (NGOs), and private firms have been developed in the last decade or so to monitor, assist, and evaluate the performance of manufacturers and retailers in sustainability. Several standards have been developed to provide guidelines supporting the three

pillars of sustainability. The leading role is played by the International Standards Organization (ISO). Apparel and textile manufacturers and retailers are the leading players in sustainable textile production. Consumers of fashion products also play a vital role in sustainability. Consumers can select or reject a product if it is not manufactured with the right use of energy, resources, or even labor.

5.1.1 Awareness of Public

The majority of consumers are changing their purchase preferences based on social responsibility, inclusiveness, or environmental impact. There are many consumers who want information regarding the garments or textiles they buy.

5.1.2 Enterprise Drive

There are several firms that have begun to advocate sustainable development issues, such as Stella McCartney, they continue to invest in research, cooperative development, and multi-level supervision to maximize sustainable development as much as possible [6].

5.2 Challenges

Since the coronavirus pandemic, the supply chain sector saw a huge disruption. While online sales of luxury and fashion products boomed, as the majority of the consumers preferred shopping online, the product delivery times increased in some cases as the manufacturers were located abroad. This lead also to an immense increase in global carbon emissions as reported by the Amazon sustainability report [1]. But at the same time, ASOS the fashion retailer became more sustainable and reduced its emissions by 13% in 2020. This has been possible because the company implemented efficient route planning for product deliveries, used renewable energies in their own facilities, and used lighter packaging with recycled materials [7].

5.2.1 Cost Issue

Tariff-higher than traditional material: International trading policies and tariffs can influence the supply of textiles globally. Tariffs are set by importing countries mainly to protect the domestic market. For example, India increased by 20% its import duty and tariffs on apparel and textiles [8]. The World bank distinguishes tariffs into three types [23]: Most Favored Nation (MFN) which is for import between members, for non-members the tariff is higher. Second is the Bound Tariff (BND) which is defined as the maximum MFN tariff level for a given commodity line. Lastly, the Effectively

Applied Tarrif (AHS) is used in real transactions practically across different countries. Certain developing countries may not have direct trade agreements with developed countries, which can be a disadvantage for textile exporters, even though the local conditions for textile manufacturers can be an advantage due to the lower labor costs.

Sustainable raw materials are also more expensive. One of the reasons is the costs associated with the certifications which may include information related to ISO, international standardized chemical tests for fabrics/leather, use of pesticides/insecticides on the raw materials, carbon emissions, sourcing of raw materials, environmental audits, and employee working condition audits, etc. Another factor is the investigation compared with existing products, more research funds, and human resources are required to carry out the process. Which for smaller brands can be a tedious process especially regarding the international sourcing of sustainable raw materials.

5.2.2 Awareness & Attitude

Consumer group-Knowledge gap also exists among retailers and manufacturers who think their shoppers know more than they do. As per a research survey carried out by [3] on the impact of sustainability on consumer purchasing patterns, it is found that 65% of executives believe that the consumers are aware of all the sustainable initiatives. But the 49% of the consumers reported that they are not aware of any information to verify sustainability claims of products. Additionally, 44% of consumers responded to having lower trust on sustainability claims of the companies.

Management-A perception many organizations have is that sustainability is more expensive. However, they do not realize that initiatives like waste reduction or energy efficiency will reduce their operational costs.

Post-process-insufficient well-organized institution or infrastructure to monitor and guide the post problems such as recycle, reusable, or waste disposal.

6 Top Leading Markets

China: As a major exporter of textiles, China has surpassed the European Union to occupy the first position in export volume since 2010 [28]. The Chinese textile industry appears unimpaired and has maintained continuous growth in the meantime of the pandemic chaos—Covid-19, the total export volume has taken up one-third (2010) of market share is now nearly half (2020). But the debate that arises globally is based on the combination of SDGs scores (Sustainable Development Goals), government policies, corporate surveys, and national public awareness—sustainability capability of Chinese textile industry remains a controversial issue. Although a trend toward a concerted push for sustainability shows in national-level policies, a non-disruptive and incremental green policymaking approach, always concerned about political stability and economic costs has been pursued by the Chinese authority

[13]. At the same time, the lack of forceful sectorial and local-level incentives and citizens' assembly for the SDGs leave China with a mixed track record on sustainability [28].

Green financing has been utilized to encourage sustainable development—200 green bonds worth 282 billion CNY were issued domestically and achieved the highest patents in three fields for further green technologies (I.E. recycling, water, and waste treatment), [13] but unclear fund allocation and zero national budget from the government will still bring sustainable development an unresolved present and unpredictable future especially among green supply chain management enterprises and green factories of the textile industry.

The lifecycle of textile could be described as one of the essential parts of sustainable textiles, such as raw materials, manufacture, and waste management. Chinese do have their advantage because linen is naturally sustainable, but cotton and excessive chemicals require more effort.

To summarize, the Chinese sustainable textile industry could be described as junior, and it is still an arduous trip to integrate it into the realistic approach.

6.1 European Union-mature but the Stressful Front-runner

Sustainable development is firmly rooted in European countries. It has been at the heart of European policy for a long time, which includes the long-term high-level statements, sectoral action plans, serious monitoring, public assembly, and specific budget. As a result, the entire EU allies are performing outstanding during the period. Unfortunately, although the economic capacity of most EU countries is sufficient to bear the heavy damage caused by Covid-19, the aftermath of the pandemic is still being calculated. The Covid-19 crisis has not only underlined the interconnectedness of the social, economic and environmental spheres but has also highlighted the importance of achieving the SDGs.

It is easy to find a product with a label of organic cotton or sustainable materials in EU from various luxury & fashion brands. The consequence is that the EU is the second exporting region and the first importing region of textile simultaneously while 60% by value of clothing in the EU is produced elsewhere [28, 12]. The greenwashing problem caused by transparency is difficult to be strictly supervised to some degree. The EU has its own strict raw material screening mechanism, production monitoring measures, and labor protection policies. However, according to the example of the sustainable pioneer brand *Stella McCartney*, it is not difficult to find out that even a brand that is extremely seeking to maximize sustainable development as much as possible, cannot monitor regularly different processes like the raw material production, raw material processing and raw material manufacturing [16]. Therefore, although EU companies can implement their own sustainable development strategies as much as possible, real sustainability is open to question if the international cooperation for transparency does not increase.

6.2 India-determiner of Future Trends

The textile industry occupies an important status in India, but the high tariffs and the increasing prices of raw materials are hard to attract more companies with advanced sustainable development technology to cooperate when compared with competitors. Followed by the manufacturing issues, India has 1,254 Global Organic Textile Standard (GOTS) certified facilities, the largest number in any single country (China only has 220), and 449 Organic Cotton Standard (OCS) certified producers. But the opposite situation also exists; for example, in Rajasthan, India, the textile industry produces highly hazardous waste [25], and proper disposal facilities are not available.

As per a report by YouGov [2], a consumer and data insights consultancy, Indian fashion designers are mirroring the sustainable fashion preferences of Indian buyers. The survey found that Indian consumers are highly motivated by sustainable fashion brands if their certifications and labels are clear and offer competitive pricing. The consumers also reported to prefer more brand communication about the sustainable fashion benefits and find it more favorable when endorsed by celebrities, along with receiving reward points/incentives for making such buying decisions. Celebrity endorsement influence for sustainable fashion was also evidenced during the pandemic when they promoted wearing masks made with older kurtas, blouses, and saris [21]. Indian consumers in the past have also used several methods of recycling and upcycling fashion textile materials. However, assessing the entire textile industry scale, sustainability is still in its infancy due to the high number of production units and lower efforts made in the direction of waste minimization, among other factors.

7 Philosophies of Sustainability

Climate and environment were supposed to have a "great year" in 2020. The need to define a new path for the world to reach the 2030 targets of "no net loss of nature" and "reducing climate change to 1.5 degrees Celsius" remains urgent. The unprecedented combination of the Covid-19 pandemic and a focus on ending racism has demonstrated our ability to act as a global society. The fact that the textile fashion industry has a substantial environmental impact on water, chemicals, and fossil fuels, and influences so many other elements and subtleties of our planet's resources, is frequently perplexing. The sector is also notorious in terms of global greenhouse gas emissions, using more energy than the aviation and shipping industries combined. During the pandemic, internet efforts and digitization have been important concerns, with significant players such as the Copenhagen Fashion Summit pivoting to develop new content using their online platform.

A sustainable business plan strives to have a beneficial impact on one or both of these areas, thereby contributing to the solution of some of the world's most important issues.

Sustainability does not imply foregoing income or putting success on hold. Instead, it has evolved into a critical component of any successful organization's strategy. Profitability, expansion, and employee retention are all negatively impacted by a company's failure to consider sustainability issues. When designing corporate strategy, taking a values-driven approach might be critical to long-term success. Many of today's businesses have adopted the triple bottom line, sometimes known as the "three Ps," which stands for "people, planet, and profit".

To mention a few, the Responsible Business Coalition, Global Fashion Agenda, Textile Exchange, Sustainable Apparel Coalition, Apparel Impact Institute, and ZDHC all exist to assist firms on their way [19]. The true difficulty will be in making executive decisions that emphasize long-term resilience over short-term profits and then enabling resources for day-to-day implementation. Many businesses are beginning to invest in solutions that will help them connect and engage their employees. Kno Global is one company that collects real-time data on worker well-being in factories to support productive workplaces, community, and social sustainability [14]. As the business case becomes clearer, more resources should be allocated.

This will have major ramifications for the fashion and garment sector, not just in terms of production, but also across the value chain, like online shopping, banking, and bill payments—digitization necessitates increasingly complicated and strong technology.

Companies should position themselves as global leaders, rather than just business leaders, by integrating their business and sustainability initiatives with the Global Goals—or, more radically, restructuring their business models—and recast their triumphs as wins for the globe. Not only does business hold the key to achieving the SDGs in the long run, but the SDGs will also assist to drive business change.

Robert Skinner, UNOP's Executive Director, emphasized the necessity of global involvement in the private sector and other stakeholders in accomplishing the SDGs [18]. Detlef Braun, a member of Messe Frankfurt's executive board, stated that since the importance of sustainability has grown in the global textile sector, Messe Frankfurt has been following this trend with its global textile events under the Texpertise Network for more than 10 years. "As a result, it is reasonable that the Sustainable Development Goals be included into our global textile events to raise acute awareness of the importance of sustainability in the textile industry," he says. The goals will be presented at the more than 50 textile events organized by Messe Frankfurt at venues around the world. Current planning includes interactive information stands, presentations, discussion forums, fair tours, and the integration of special activities in the trade fair program.

8 Alignment for the Future

Despite, sustainability having been discussed persistently in recent years, fewer actions have been taken on it. The textile industry, as one of the biggest polluters in

Short Term	Medium Term	Long Term
Turn consumers' environmental awareness into sustainable action	Innovation in fibres and processing	New textile economy
	Transparency & supervision	
	Worker-driven initiatives	

Fig. 1 Time frames of sustainable strategies

the world, the urgency of transition is inevitable. Companies and brands should take the responsibility for a better and greener future.

Here are several tactics that key players can apply to foster sustainability in the textile industry within three timescales.

Here are several tactics that key players can apply to foster sustainability in the textile industry [5]. As a new textile economy is expected to be developed eventually, therefore, to achieve this long-term vision, short term and medium term are designed, separating strategies into phases to be more executable (Fig. 1).

8.1   Short Term

What practitioners in the luxury and fashion industry can do for the short term is to reinforce consumers' positive attitudes toward sustainability and encourage them to follow through with sustainable behaviors. Although consumers often self-report having a positive attitude and intentions toward sustainable consumption, they don't subsequently pursue a sustainable choice and behavior. The gap between "attitude and behavior" is challenging for marketers and needs to be closed with sustainable efforts in the right directions following the SDGs and the internal policies aligned with global strategies. Comparing to fast fashion, affordable, and convenient goods, sustainable labels are less attractive due to the more efforts consumers could take—sustainable goods tend to cost more and are seen as less trendy.

To trade off convenience for eco-friendly, consumers require a persuasive reason. Hence, here listed 5 elements that companies could take into account for brand communication: social influence, habits, individual self, feelings, cognition, and tangibility [27].

Thus, sustainable activities or promotions could work on providing chances for consumers to convey a "care for the earth" impression to the society or within a group of communities. The easy way to change a behavior is to cultivate a new habit replacing the old one. For example, Levis hold the "Water less" campaign, cooperated with Water.org. Their campaign is simple and frank with the hashtag #doingyourpart, challenging people around the world to change their habits by using less water, they straight up say the fact that you don't have to wash your pants every week.

8.2 Medium Term

The demand for sustainable products will be consistently increasing because of the shift toward consumer behavior, more actions need to be followed up. As discussed earlier in this chapter, innovating new fibers and processing technologies can be seen as the initial step moving onto the medium-term plan.

The need for transparency in the industry is essential to supervise the textile system and to avoid greenwashing. Without transparency, a sustainable, accountable, and fair textile industry will be hard to achieve. The feasibility of transparency and supervision can be determined by establishing global standards with governments affiliation—regulations, and policy-led incentives, following the model of mandatory financial reports [11].

In addition to this, applying blockchain to the industry for traceability is considered as the novel strategy to the next level of transparency. This technology allows each garment to obtain a unique token which enables the company to verify every step of its production and create a digital history of that information including location data, content, and timestamps. All this information is presented to consumers via an interface they can access through their item's QR code or NFC-enabled label. With blockchain applied in the supply chain, from the use of raw materials to the last finished stamp, the whole process is traceable, therefore ensured the authenticity of sustainable contribution.

Meanwhile, the protection of workers' human rights should be considered. A garment assembly line requires many workers, starting from textiles yarns, and fabrics. The intensive workload and low-standard employment are seen as the common facts in the textile industry for a long time. However, to conduct a sustainable industry, human rights can't be ignored. Garment workers need to be treated decently with better working conditions and empower by training and education as the rapid change, especially in production technologies, negatively affects workers who have the lowest education level in the textile industry [5].

8.3 Long Term

For the long-term vision, the dynamics of the textile industry need to shift based on circular economy principles in making sure that economic development can happen in a way the planet can afford. To briefly explain the term "circular economy"—use resources sparingly and recycle endlessly [15]. With all the efforts introduced, a likelihood of a circular economy-based textile industry can be achieved. In a real sustainable world, consumers can be conscious about their consumptions of textile products, a lifecycle of a product can be traced from the supply chain ensuring fibers and chemicals used in textile production are safe for both workers and consumers to the after-sale service, extending product life by repairing and recycling. Circularity is certainly not the end goal, but an important pathway contributing to the end goal—to attain a greater human and ecological well-being.

References

1. Amazon sustainability report (2020). Further and faster together. https://sustainability.abouta mazon.com/pdfBuilderDownload?name=amazon-sustainability-2020-report
2. Bhatia, D (2019) More than 8 in 10 Indians are open to buying sustainable fashion items. https://in.yougov.com/en-hi/news/2019/08/06/more-8-10-indians-are-open-buying-sustainable-fash/
3. Capgemini (2020) How sustainability is fundamentally changing consumer preferences. https://www.capgemini.com/wp-content/uploads/2020/07/20-06_9880_Sustainability-in-CPR_Final_Web-1.pdf
4. Cernansky, R (2021) Is there space in sustainable fashion for synthetic fabrics? https://www.voguebusiness.com/sustainability/is-there-space-in-sustainable-fashion-for-synthetic-fabrics
5. Daheim C, Nosarzewski K, Kołos, N (2019) The future of sustainability in the fashion industry. https://www.laudesfoundation.org/en/resources/future-sustainability-fas hion-industry-delphi-final-report-futureimpacts-ca-2019-v7.pdf
6. Edie newsroom (2018). Stella McCartney: Policy must encourage businesses to make fashion sustainable. https://www.edie.net/news/7/Stella-McCartney--Policy-must-encourage-businesses-to-make-fashion-sustainable/
7. Edie newsroom (2021). Asos reduced emissions by 13% in 2020, despite pandemic parcel boom. https://www.edie.net/news/6/Asos-reduced-emissions-by-13--in-2020--despite-pandemic-parcel-boom/
8. Fibre2fashion (2018). India increases import duties on textiles & apparel. https://in.fashionne twork.com/news/India-increases-import-duties-on-textiles-apparel,998359.html
9. Freitag (2021) From truck till bag. https://www.freitag.ch/en/production
10. Roach Y (2017). What is the difference between natural, synthetic & semi-synthetic materials? https://www.undershirts.co.uk/blogs/research/basics-the-difference-bet ween-natural-and-synthetic-textiles
11. George S (2018) Stella McCartney: Policy must encourage businesses to make fashion sustainable. https://www.edie.net/news/7/Stella-McCartney--Policy-must-encourage-businesses-to-make-fashion-sustainable/
12. GmbH M-B (2021) The European market potential for recycled fashion. https://www.cbi.eu/market-information/apparel/recycled-fashion/market-potential
13. Holzmann A, Grünberg N (2021) "Greening" China: An analysis of Beijing's sustainable development strategies. https://merics.org/en/report/greening-china-analysis-beijings-sustainable-development-strategies
14. Knoglobal (2021) https://www.knoglobal.com
15. Kunzig R (2020) Is a world without trash possible? https://www.nationalgeographic.com/mag azine/article/how-a-circular-economy-could-save-the-world-feature
16. McCartney S (2016) Transparency In the supply chain and modern slavery statement p. 9
17. Ramchandani M, Coste-Maniere I (2020) Leather in the age of sustainability: A norm or merely a cherry on top?. In: Muthu S. (eds) Leather and footwear sustainability. Textile Science and Clothing Technology. Springer, Singapore. https://doi.org/10.1007/978-981-15-6296-9_2
18. Texintel (2019). Messe frankfurt and the U.N. to work together promoting sustainability in the textile industry. https://www.texintel.com/eco-news/messe-frankfurt-and-the-un-to-work-together-promoting-sustainability-in-the-textile-industry
19. Textileexchange (2020) Leading NGOs unite as the FASHION CONVENERS to accelerate sustainable transformation of the apparel and accessory industry. https://textileexchange.

org/leading-ngos-unite-as-the-fashion-conveners-to-accelerate-sustainable-transformation-of-the-apparel-and-accessory-industry-2/

21. Tikoo R (2021) Making sustainability fashionable. https://timesofindia.indiatimes.com/blogs/development-chaupal/making-sustainability-fashionable/
23. Setyorini D, & Budiono B (2020) The impact of Tariff and imported raw materials on textile and clothing export: Evidence from the United States market
24. Stand.earth (2021) https://fashion.stand.earth/key-findings/fashions-fossil-fuel-problem
25. UNFCCC (2021) Sustainable textiles for sustainable development—India. https://unfccc.int/climate-action/momentum-for-change/activity-database/sustainable-textiles-for-sustainable-development
26. Villena VH, Gioia DA (2020) A more sustainable supply chain. https://hbr.org/2020/03/a-more-sustainable-supply-chain
27. White K, Habib R (2018) SHIFT—A review and framework for encouraging environmentally sustainable consumer behaviour. Sitra, p. 11
28. World Trade Organization (2021) World trade statistical. World Trade Statistical Review :55

Consumer Attitudes Toward Sustainability in the Garment Industry—A Consumer Study in Hong Kong

Si Kei Isabella Ng and Cecilia Mark-Herbert

Abstract Sustainable development in the textile industry is intimately connected to consumer behavior. This chapter concerns consumers behavior in Hong Kong, focusing on the purchasing of garments. Social practice theory offered an understanding of consumer behavior that goes beyond attitudes in the studied focus groups. The results indicate that the complexity in the textile industry contributes to difficulties to capture sustainability aspects in purchasing a garment for the consumer. Therefore, efforts need to be made to educate consumers, which will enable them to be part of continued sustainability transformations of the textile industry.

Keywords Communication · Competence · Fast-moving consumer goods · Meanings · Materials · Textile · Social practice theory · Value chain

1 Introduction: Sustainability Challenges in the Textile Industry

1.1 Needs for Alternatives to "Fast Fashion"

There has been a vast number of promotions and advertisements on online platforms which lead human beings to purchase unnecessary garments throughout the past decade [28]. A term called "fast fashion" has been widely used, which reveals the consumer demands for cheap and seasonal textile products which deteriorate the environment [38]. The consumption patterns are determined by various context-bound factors [31], which are associated with financial, social, and environmental awareness of sustainable development implications [9].

S. K. I. Ng (✉)
Department of Earth Sciences, Uppsala University, 752 36, Uppsala Uppsala, Sweden

C. Mark-Herbert
Department of Forest Economics, Swedish University of Agricultural Sciences, 750 07, Uppsala Uppsala, Sweden
e-mail: cecilia.mark-herbert@slu.se

© The Author(s), under exclusive license to Springer Nature Singapore Pte Ltd. 2022
S. S. Muthu (ed.), *Sustainable Approaches in Textiles and Fashion*,
Sustainable Textiles: Production, Processing, Manufacturing & Chemistry,
https://doi.org/10.1007/978-981-19-0874-3_2

[38] state that the textile industry is the second largest polluter in the world. A significant number of challenges in sustainability areas, including resources use, pollution, and exploitation, are identified during the production of garments. Hence, this leads to high levels of greenhouse gas emissions in different production phases in this industry. The manufacturing process, such as fiber and yarn production, as well as garment disposal, are contributing to the pollution [32]. Moreover, fertilizers and pesticides are needed in growing cotton [25]. This example shows that a tremendous amount of water, energy, and chemicals are required in various stages of production, use, and disposal. When it comes to the social aspect, [49] indicate that the outsourcing of apparel production points to needs for global perspectives, which raises issues of people's awareness toward the safety issues in manufacturing factories and labor exploitation.

Green fashion, eco-fashion, and sustainable fashion these concepts represent understandings of the production of fashion apparel that should not negatively affect society or the environment [44]. These terms include both production, distribution, sales, use, and waste management aspects of a circular system. It includes, according to [50], perspectives of consumers an "after-life" can be granted for apparels to reduce the waste, such as second-hand stores donations or cloth swapping. Moreover, regarding the perspectives of fashion enterprises, organic fiber or recycled materials can be adopted for production and fair working conditions can be established, all of which are transparent and traceable [3]. As a result, there has been a rise in long-term actions regarding system development for sustainability in the fashion industry.

In the recent decades, sustainably consciousness and environmental awareness have been enhanced in many societies [44], and therefore, the apparel industry has engaged in responsibilities toward sustainable development [28]. There is no doubt that consumers are expecting more sustainable implementations, and this has intensified garment companies to secure their business in line with related policies [5]. Asian cities suffer from a rapidly diminished landfill area due to serious air pollution and a high level of garbage disposal [11]. Therefore, the problems brought by the fashion supply chain become more visible, and concerns toward "green" issues have arisen among consumers. These perceptions and values are likely to influence the consumer behavior of garment. [31] also demonstrates that garment retailers have advertised durable products with higher opportunities to be recycled to show their efforts to strive for sustainable development. As the brands may earn profits and reputations by introducing sustainable practices, there are ongoing steps with more enterprises to expand their sustainable apparel production [44].

1.2 Consumer Perspectives: Challenges for Transformation

Consumer understandings and practices are central to a sustainability transformation. Since consumers carry different perceptions toward sustainability of fashion brands, studies on the relationship between consumer attitudes and sustainability have been made in various countries and cultural contexts [4]. However, previous studies have primarily focused on countries other than Hong Kong. Consumers in Hong Kong are less likely to recognize the environmental harm caused by purchasing behavior, compared to the Western countries [31]. This may be due to insufficient sustainability awareness concerning the garment industry. Moreover, there is a lacking incentive systems that may justify sustainability efforts (legislation, taxation, subsidies, etc.). Therefore, this research project has been concentrating on investigating consumer perceptions and behaviors in relation to sustainability for garment purchasing decisions in Hong Kong.

In order to increase the awareness of consumers toward sustainable fashion, it is essential for fashion brands to develop and advertise their sustainable measures [30]. However, it remains questionable whether these measures positively influence consumer behavior in buying clothes after these brands take sustainable measures [24].

1.3 Aim and Delimitations

This study focuses on how Hong Kong citizens perceive sustainability and whether it is an element to modify their purchasing decision. The aim of this research project is to determine the key factors, which affect consumer behavior and attitudes in Hong Kong in terms of sustainable development in the garment industry.

The following research questions are key for understanding consumer perspectives of sustainable development in the apparel industry:

1. What are the factors that affect consumers in purchasing clothes?
2. What are the perceptions of consumers regarding sustainability in the garment industry?

Delimitations are made with an empirical focus on current consumer perspectives in Hong Kong. From a theoretical perspective, the Social Practice Theory (SPT) offers a focus on practices rather than attitudes and consumer profiles.

1.4 Outline and Ambitions

The presentation of this chapter follows the traditions of an empirically driven research project. Starting with a presentation of a theoretical perspective, followed by an account of the approach, and thereafter a presentation of the results. These results are put in relation to other contemporary studies that serve as dialogue partners. Lastly, conclusions are presented along with reflections on policy implications.

2 Theoretical Framework

2.1 Sustainability Issues in Garment Industry

There are a significant number of sustainability issues in terms of the garment industry, especially in the developing countries where most of the textile production occur. The major perspectives include manufacturing process, transportation, waste disposal, and labor employment, as demonstrated in Fig. 1.

Starting with the manufacturing process of natural and synthetic fibers in Fig. 1, [25] explain that a tremendous amount of water, energy, and chemicals are required in this process, for example, a lot of fertilizers and pesticides are used in growing cotton and other fiber crops. This leads to several ecological upsets such as water pollution, decrease in soil fertility, and loss of biodiversity [44]. Water shortages and environmental disasters can be caused by intensive crops growing for the textile industry, for example, the Aral Sea has shrunk over 90% as water has been diverted for cotton production [37]. Toxic chemicals are also applied in conventional cotton agriculture and harvesting. The synthetic fibers, for example, nylon and polyester, are made from finite resources, which in itself is a sustainability challenge.

The transportation process is also associated with numerous system-bound challenges. Globalization of the textile supply chain has led to the situation that materials are grown, fabric production factories, garment production factories, and the final retailer for selling are all situated in different geographical locations [33]. There is research that investigated that a pair of Lee Cooper jeans could take a global journey with over forty thousand miles [6]. Therefore, carbon emissions cause a rise of awareness and concern toward "garment miles" [38].

Regarding disposal of clothing, [38] claim that there is approximately 1.8 million tons of garment waste every year. While some of these volumes have been collected and exported for recycling or further usage, most of them have been sent to landfills as they do not fit the purposes and requirements [25]. Moreover, in order to keep the local textile industries active in the developing countries, sending second-hand garments to those nations is not suggested [33].

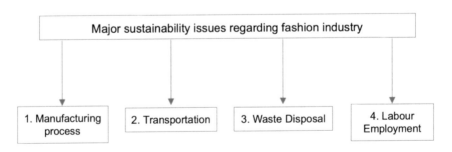

Fig. 1 An overview of the major sustainability issues regarding the fashion industry

Furthermore, there are concerns about labor treatment in garment factories [14]. Due to the price drop of garments, exploitation of workers may arise that leads to risk for workers not being able to afford daily necessities for themselves and their families [43]. Therefore, they may have to work excessive hours but only to a minimum wage. Some consumers boycott fashion brands who exploit workers in developing countries, however, some unintended disadvantages may cause this action to be ineffective, such as cancelation of job contracts that make the workers lose their jobs [44].

2.2 Consumer Behavior

When consumers are making a decision, their behaviors are affected by both psychological and sociological factors [19]. Psychological components are regarded as internal considerations, including motives, memory, attitudes, personality, and learning. Sociological elements are external factors, for example, reference groups, culture, social status, and marketing activities [2]. In order to understand the conditions for consumer markets, researchers and markets in an enterprise consider these factors.

2.2.1 Psychological Considerations

Needs and wants are distinguished in consumers' motivations during purchasing decisions. Maslow's Hierarchy of Need (Fig. 2) is widely recognized and commonly used in illustrating a theory of human needs. There will be a progression to the higher level when the needs are fulfilled on a lower level [36]. Garments can fulfill the needs in all the levels within this hierarchy as garments fulfill basic human needs by serving for three purposes, including protection, decoration, and modesty [35].

More than just basic needs, garments can be a way to reflect one's individual identity, determining psychological needs and self-fulfillment, as shown at the top part of the hierarchy model in Fig. 2, [33]. According to [45], the actual self, the social self, and the ideal self, are the three main aspects to perceive self-concept: the actual self is how one view himself or herself, the social self is how one view other people perceive him or her, and the ideal self is how one would like to be perceived by others. This self-concept is applied in the social identity theory, where individual characteristics such as abilities, interests, and potentials are included [45].

In general, the consumer behavior can be reflected by one's attitude toward himself or herself [2, 19] report that this is especially applicable toward young people, as they are more eager to relate self-expression to their own identities in order to develop belongingness and social status.

There are various internal reasons that can drive consumers to purchase goods, both rational and emotional factors are essential motives in terms of consumer behavior [2]. The general selection of goods, such as price and quality, are rational

Fig. 2 A classical model for understanding human needs based on Maslow's Hierarchy of need [36]

factors, while emotional ones are likely the uncontrollable feelings that satisfy an individual's needs and wants during the decision-making process (*ibid.*). It is common for consumers to buy products for their values instead of their functions, therefore, the brand name, the product image, and the after-sale service are regarded as motives [35]. Consumer loyalty toward the company is further developed if the goods and services can meet the goal of the consumer. It is worth mentioning that consumer behavior is not affected by concrete rules of internal factors, and each purchasing decision may satisfy more than one need [19].

2.2.2 Sociological Considerations

Regarding the external sociological aspect, peer groups and social groups can significantly influence one's aspirations and behavior, which can be in the groups of formal or informal [33]. As explained in the psychological elements above, consumers can develop personal identities in purchasing and consuming products, mainly in the social relations area. Thus, social status can influence consumer attitudes, which different purchasing patterns can be shown from different class, and income levels [2, 19] state that people from higher social status and income level have a higher willingness to purchase as they have more unrestricted capital.

Furthermore, culture is also a major reason that can determine consumer behavior. This is a complex whole acquired by human beings that contains knowledge, belief, morals, and other habits in a society [19]. People are not likely to be aware of culture as it naturally takes place and affects various aspects of how people behave [2]. Also, culture does not interpret and specify appropriate behavior, thus, it explains the way of thinking and acting of the majority in the society [44]. There are subcultures within certain groups in a country, such as ethnic and peer groups, which can all affect consumer behavior [13].

2.3 Ethical Purchasing

Much research regarding consumer behavior is based on consumers' perspective as an "economic man" (*homo-economicus*), which refers to a rational person that acts in accordance with his or her own needs and interests [23, 39] claim that there is a gap between the intention and the actual behavior concerning this classical view of consumerism. Consumers argue that they are both ethical and sustainable in many cases, however, they rarely consider these factors during a decision-making process [39]. Due to the difficulty of considering ethical issues in daily life, the number of sustainable decisions made is less than the expected ones based on data (*ibid.*). Although ethical intentions can be suggested as reliable, they do not always draw the actual results as both psychological and sociological factors affect the real-time decisions [19]. As a result, these factors must be aligned with ethical and sustainable knowledge in order to train ethical consumer behavior to become a habit [39]. The patterns and dimensions of ethical consumption must be planned and built before they develop into a habit, hence, consumers can consistently behave ethically in the purchasing decisions [8]. In conclusion, more than just the responsibilities of the brands in making sustainable information available and transparent, consumers also have the responsibility in developing an ethical lifestyle in order to minimize the gap between intention and behavior.

2.4 Social Practice Theory

There are numerous theories that can be utilized for determining and explaining human behavior, and the ontological starting point can be used as a social order which has two opposite directions. These foundations for evaluating human behavior rest on *homo-economicus* or *homo-sociologicus* [41]. *Homo-economicus* claims that individuals' rational intentions shape the social order in the world based on psychological individual needs, while *homo-sociologicus* suggests a more sociological perspective that individuals follow the rules and norms which are commonly accepted (*ibid.*).

Social Practice Theory (**SPT**) is a new way for theoretical perspective. This perspective gives more attention to what consumers do regarding practices in purchasing, use, and disposal of products. rather than focusing on consumer attitudes. SPT is an approach that combines the components in daily routines in order to understand the actions of human beings [1], hence, consumers' actions instead of the attitudes are identified. It is further proposed that a conceptualization of consumers as the active practitioners rather than the users [40]. These practices are more than the actual human behavior, which can be observed by researchers as they require what people say and do in such performance (*ibid.*). They are usually routine activities of human beings such as eating and sleeping, and people who participate in these activities are all engaged in the new form of analysis [41].

There is a great deal of perspectives existing in terms of the components and the role in SPT [41]. Hence, according to [40], SPT is explained with a practical theme based on consumption, which makes the theory easier to be applied. A model developed by [48] is presented in Fig. 3. It identifies the major elements in social practices of consumption, and there are three mandatory components included, which are materials, meanings, and competence. They are not isolated but interconnected, thus, they are combined to strike a balance in the actual practices [48].

Materials include the actual physical objects that are tangible, for example, human bodies, tools, infrastructure, and technologies, and they are usually directly related to human beings' daily lives [40]. *Meanings* refer to the beliefs and understandings which are socially shared and connected with respect to the materials, different associations, and culture may have different senses of what meanings are appropriate [48]. *Competences* encompass techniques and skills that are required to perform a practice, thus, the way to recognize and respond to a certain behavior [40]. Practices are developed from continuous connections and linkages of its components, and

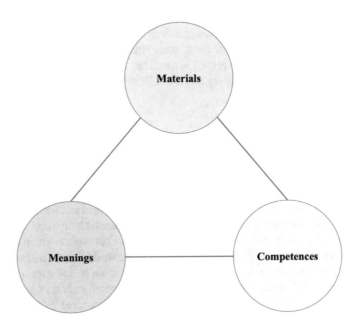

Fig. 3 Components of social practice to explain behavior [48: 14]

the change of the components leads to change of the practice (*ibid.*). People tend to continue engaging in and carrying on the practice when they gain positive experiences by performing the practice (*ibid.*).

2.5 A Conceptual Framework

The conceptual framework serves as a synthesis of the theoretical framework, thus, this is applied in the discussion later on to explain the consumer behavior in buying garments. A detailed conceptual model is developed for this study with the Social Practice Theory (**SPT**) shown in Fig. 3 and text Sect. "Social Practice Theory". The three components in Social Practice Theory may be crucial in affecting consumers to make sustainable decisions and choices, which are embedded as actions in sustainable fashion consumption practices, as indicated in Fig. 4.

Consumers have certain actions in purchasing garments in this practice, including sustainability considerations. Regarding the first element, materials is about the activities involved in the decision-making process of garment purchasing. The second element, meanings, is the factors and issues associated with the decision, while the third element, competences, is the knowledge and awareness of consumers toward sustainability of a garment and the industry.

2.5.1 Conceptual Framework: Consumer Perceptions

Consumer behavior and perceptions are context-bound. They may change over time, which means that the actions and the practice of purchasing garments change. Based on the conceptual framework and the background about consumer perceptions in the garment industry, Fig. 5 shows the summary of the connections between consumer awareness and purchasing intentions. Hence, consumer behavior can be influenced by the knowledge and factors toward sustainability issues.

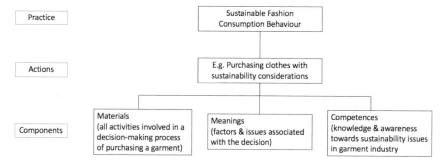

Fig. 4 Conceptual framework underlying, based on the elements of practice according to [48: 14] with modifications

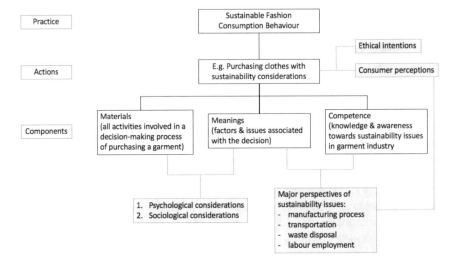

Fig. 5 Combination of empirical studies and the conceptual framework, based on the elements of practice according to [48: 14] with minor modifications

The above figure reveals a detailed relation between consumer perceptions and the conceptual framework. [29] investigate to what degree consumers are more willing to pay more for products from brands that are environmentally responsible if they have stronger attitudes toward environmental issues. This suggests that consumers are more likely to pay attention to environmental policies implemented by companies before a purchasing decision if they have more knowledge about these issues [19]. However, there are no fundamental modifications in consumer behavior of sustainable fashion, and this may be due to the complexity of this field, with individual and social factors as well as barriers [33]. As a result, consumers may have many other factors and priorities in choosing their garments with a rise in sustainability concerns and awareness. Moreover, [46] argues that there is a comparatively higher level of concern toward labor issues and a comparatively lower level of awareness toward the environmental consequences brought by intensive cotton production and synthetic fibers. This may be due to a lack of media reports and coverage about the manufacturing process of textile products [15].

3   Method

3.1   Research Approach and Data Collection

The approach reflects an empirically driven research design, where a literature review served as a background for making theoretical and methodological choices. Based on these understandings, two focus group sessions were conducted. According to [12] two focus groups are seen as the minimum number for a research project. Due to the Covid-19 pandemic situation, data were collected from two groups.

A focus group dialogue allows for sharing a great variety of ideas and thoughts, which can explore consumer perspectives and understand the actions of both individuals and groups. [10] reveal that during the process of focus group sessions, all words including body language, tone of voices, and reactions can be recorded instead of statistics. In this project, the subjective assessment was an ideal way to join the respondents together and to hear their views when they practice the garment purchasing decision.

Potential participants in the focus group were identified as Hong Kong citizens born in the 1990s. This group is a dynamic and consumption-oriented generation, and they are also more conscious toward sustainability than other previous generations as they have been educated to be more environmentally friendly [17]. As a result, the young 90 s consumers are assumed to be more willing to act sustainably to make positive contributions to the environment (*ibid.*). The discussions toward sustainable consumption are more effective and informative by setting the most conscious generation as the target group. Therefore, both men and women ranging from 20 to 30 years of age were targeted.

In terms of the procedure of recruiting participants, the number of 2 to 12 participants can be engaged in a focus group session in which a larger sample with more opinions can be obtained [10]. Moreover, [7] illustrate that a group size of four to six people is regarded as ideal, where all subjects can share and explain their own thoughts and experiences in detail in a less intimidating environment. In this research project, four respondents were initially invited to participate in one of the focus group discussions (Table 1).

The first focus group conducts by inviting participants who know each other, while the respondents from the second focus group were selected on social media, which is a request that participants do not know each other beforehand. The two approaches in recruiting persons to engage in the focus group were made to learn about the willingness to express opinions in a group with individuals that know each

Table 1 Information of the participants in Focus group 1 and Focus group 2

Name	Gender	Age	Education level	Focus group
Vincent	M	22	Bachelor	1
Tiffany	F	22	Bachelor	1
Alex	M	26	Bachelor	1
Christy	F	24	Bachelor	1
Makalya	F	22	Master	2
Kelly	F	21	Bachelor	2
Jane	F	28	Bachelor	2
Hilary	F	23	Master	2

other versus a group that does not. According to [10] data collected from a focus group with members who know each other is likely to be more in-depth as they have a sense of belongingness and feel safer to share personal consumer behavior, as well as presenting in a less serious discussion.

A focus group session generally has a time frame below 2 h [7]. Once the focus group session is arranged, the group facilitator or the moderator plays an important role in giving a clear explanation of the purpose, facilitating interactions between the participants, and helping them to feel at ease [27]. During the discussion, the moderator is responsible for notes taking and observation [10], and the moderator is the researcher of this study during the session. In this project, the focus group dialogues were conducted via Zoom, where all of us could clearly see and hear each other. An interview guide is provided in Appendix 1.

Focus Group 1 included four participants (age 22–26) and one moderator. The two men and two women all know each other. The session was scheduled on a Saturday (25 April, 2020) night at 21:00 Hong Kong Time (HKT, UTC +8) and the discussion started at 21:30. The half-hour prior to the discussion was to let the participants chat a little bit to clear the air before discussing the themes about the study. The subject-related discussion lasted for 1 h and 6 min.

Focus Group 2 included four participants (21–28) and one moderator as well. The participants, all women, did not know each other. The session was scheduled on a Sunday (26 April, 2020) afternoon at 15:00 Hong Kong Time (HKT, UTC + 8) and the discussion started at 15:15. The participants got the opportunity to know about each other during the 15 min prior to the start of the actual session of the research. The subject-related discussion lasted for 1 h and 3 min.

3.2 Data Analysis and Quality Assurance

As stated by [22], qualitative analysis is different from quantitative research method as data collected is words in a text instead of statistical numbers. As qualitative data analysis is context-bound, the results are not entirely predictable [12]. The focus group discussions were audiotaped and then transcribed, this refers to a translation of the verbal language into text. The process of listening to the conversations in the recording is then followed by analyzing the content in chronological order according to the discussion flow [26]. The empirical results section is strengthened by some selected quotations. The questions and themes for the sessions were based on the theoretical framework, and the results were discussed in terms of the three elements in Social Practice Theory, which are materials, meanings, and competence (see 2.4 Social Practice Theory.

In terms of research quality, it is crucial to strengthen the trustworthiness of the research; hence, transferability and credibility are the most significant methods for the evaluation [34]. The information shown should be traceable to identify the aim and also in a logical way, which is based on numerous perceptions from previous studies to increase the strength of the research [10]. A wide selection of previous

Table 2 Methods to ensure validity and reliability (based on [42: 78–79]), with modifications

Research design tests	Example of related techniques	Applied in this study
Validity	Construct chain of evidence in collecting data	Focus group discussions are transcribed
	Adopt diagrams to assist explanation of the concepts and findings	Graphic models are used to review the framework which explains the results
	Third-party review of evidence	Professional and skilled proof reading by the supervisor of this study
Reliability	Mechanically data recording	Focus group discussions are taped
	Record actions and observations as concrete as possible	Focus group sessions are recorded, notes jotted based on observations
	Provide full account of theories and ideas	Follow the interview guide established in appendix 1

research work has been chosen as references for supporting the theories and empirical data in order to achieve transferability. The target participants who were born in the 1990s were selected as they are consumption-oriented and dynamic with the highest awareness toward sustainable development [17]. This was decided by the researcher in order to lead the study into higher relevance and more time efficiency. [10] claim that credibility is to ensure the connections of the categories and also the observations, conclusions drawn by the reader should be the same as the researcher. Techniques are established to strengthen the validity and the reliability [42], which are also important in research (Table 2).

Table 2 shows that diverse techniques were introduced to ensure both the validity and reliability of this study. The methods applied in the research are presented in the data analysis, which was entirely conducted to assure the quality of this study.

There are ethical considerations in terms of focus groups and most of the social research [20]. Researchers have to be honest and inform the invited participants about the aim of the research and the purposes of their contributions, as well as not pressure them to talk [16]. The moderators have to specify the contributions of each participant and what they are going to share with others as there is more than one participant within the group [10]. All respondents engaged in the discussions are encouraged to keep the opinions confidential, moreover, researchers take the responsibility of anonymizing the data collected. Also, the participants were all asked to sign a GDPR form, which covers informed consent and the data legislation (Appendix 2). It is crucial to ensure that the information about handling sensitive material is given to all of them.

4   Results and Analysis

The results are presented in accordance with the themes in SPT, materials, meanings and competences.

4.1   Materials

Materials are regarded as the first component, which involves the activities during the decision-making process of purchasing a garment. In this section, the plans for buying garments and the motivations for consumption are drawn in the analysis. These topics can answer the first research question: *What are the factors that affect consumers in purchasing clothes?* This led to the discussion of the properties which the participants look for in the garment they buy.

Both focus groups proposed that emotions could be the motivations in affecting buying decisions. As the mood of the day may drive them to purchase clothes on impulse, there were no consumption plans for this reason. Kelly from Focus Group 2 and Tiffany from Focus Group 1 suggested that negative emotions, such as unhappiness and stress, motivated them to go shopping:

Kelly: "Hong Kong people are always under pressure, so I agree that many of us choose to buy clothes that we want in order to feel happy and relieved. We earn money in our jobs, both full- time and part-time, so we have to use the salary to buy apparels that we love instantly, a way to compensate for our efforts."

Tiffany: "Yes, that's really sad. Also, my mood on that day is important regarding my purchasing decision. If I am unhappy or stressed, these negative feelings can definitely drive me to buy things."

These quotes could be linked to the emotional factors of psychological considerations within the theoretical framework. These are uncontrollable feelings that make an individual's needs and wants satisfied in the decision-making process [2]. Moreover, the struggle between needs and wants was mentioned in Focus Group 2. During the decision-making process, Jane illustrated that a big sale could motivate her to purchase clothes. Makayla replied that a discount could also be a motivation, and Jane agreed with that:

Jane: "Yes, especially during sales, I would think that I would regret afterwards if I didn't buy what I wanted."

Makayla: "It's really hard to resist if there's a huge discount price. When there's a seasonal change, many retail stores offer a big sale, and that motivates me to purchase a lot of clothes. I also like to walk around outlets to see if there are cheap clothes with high quality. I love to fill my closet with many clothes."

Jane: "Same here, a discount or a coupon can drive me to purchase clothes, but actually I may not need them. I just want to own them at that moment."

As indicated in the theoretical framework chapter, there will be a progression to the higher level when the needs are fulfilled on a lower level [36]. More than just basic needs, the participants purchase apparels that they want to achieve self-fulfillment. Therefore, generally speaking, garments can fulfill the needs in all the levels within this hierarchy. According to [45], the actual self, the social self, and the ideal self, are the three main aspects to view self-concept: the actual self is how one views himself or herself, the social self is how one perceives other people perceive him or her, and the ideal self is how one would like to be perceived. Christy and Tiffany from Focus Group 1, and Hilary and Kelly from Focus Group 2, explained that self-concept was an important motivation in purchasing clothes:

Christy: "I usually go to ask others if the garment suits me before buying it. This is an important consideration."
Tiffany: "So they are my considerations too as I care how others perceive the garment very much. I hope others think that it's beautiful too."
Hilary: "But then you can't try the apparel if you buy it online. I prefer to shop physical stores and try the garment as I want to see how it really looks on me before buying it. If I hang out with friends, I can also ask for their opinions to see whether the outfit suits me well. Others' suggestions are quite important to me during a decision-making process."
Kelly: "I'm quite easy to be affected by others. My friends ask me to buy it as it fits me well, then I'll be easily motivated and purchase that."

These proved that they would like to ensure that the garment fitted on them by asking for opinions from others. The decision could be affected by the perceptions of other people. As a result, others' attitudes toward themselves can reflect the consumption behavior [2].

In addition, regarding the garment category for different occasions, the purchasing plans could be involved in the decision-making processes. This can be regarded as a culture factor and be classified into the sociological considerations in the chapter of the theoretical framework. Culture does not represent appropriate behavior, hence, it shows the way of thinking and acting of the majority in this society [44]. The behavior can be affected by different social groups or activities. During the discussion of Focus Group 1, Tiffany and Christy interpreted that special occasions motivated them to buy particular outfits:

Tiffany: "Same as you, I don't have many special outfits, just remember that I dressed up in last Halloween party with a bizarre outfit because I have to integrate into that social activity."
Christy: "Oh yes, only when there're these kinds of parties, I planned to buy something special which I won't wear in normal situations. Also, in occasions like high-table or graduation dinner, I look for particular garments that fit."

Furthermore, culture can influence a number of aspects of how people behave, however, they are not likely to be aware of culture as it naturally takes place [2]. Jane and Hilary specified the purchasing decisions between outfits for the workplace and the casual wear:

Jane: "I have been working for several years, compared to my consumption behavior during university, I buy less clothes now, especially shorts, because I know that I can't wear them during my work. Therefore, I'd consider this factor before buying clothes, and I'm more aware of the quality instead of the design now."

Hilary: "I totally agree that as I started working last year, I find that I buy less casual wear. The clothes which I wear in the workplace are somehow more "high-class" as I have to respect my job position and to show my status that I engage in this social circle."

Jane: "And then I find myself naturally influenced by my colleagues and we wear more and more similar clothes."

Hilary: "Yes, and my casual wear is more like "myself". I bought cheaper clothes when I was a student as I don't earn so much in the part-time job. The difference is now I can afford to buy casual wear that is more expensive with higher quality."

It is believed that there are various plans for buying different types of garments. Both focus group sessions had conversations about the fact that garments represent the social status of the wearer. The garment categories affect the properties which the participants look for. Kelly in Focus Group 2 proposed:

Kelly: "Colors and designs are important in choosing a garment. We said that quality and durability are crucial especially for winter apparels, and I think I have more summer apparels, and colors and the cutting are more important for t-shirts and skirts."

As a result, there are different motives which can drive the participants to buy a particular garment. A significant number of activities are involved during decision-making processes, including emotions, discounts, others' views, and the garment categories. These considerations can affect the plan of consumption or to purchase on impulse.

4.2 Meanings

The second element, meanings, refers to the factors and issues associated with the decision-making processes. In this section, concrete factors that affect participants in purchasing clothes were discussed, and this can continue to explain the first research question: *What are the factors that affect consumers in purchasing clothes?* Also, the views and perceptions regarding sustainability in the field of the garment industry were expressed. Hence, the second research question can be answered: *What are the perceptions of consumers regarding sustainability in the garment industry?* This part is to understand the important factors when the participants consider purchasing a garment, as well as whether the garment sustainability level is prioritized in a decision.

To begin with, psychological factors such as price, quality, durability, and materials used are indicated. As demonstrated in the theoretical framework chapter, these are rational factors that suggest the general selection of goods [2]. Both focus groups explained that price and quality were essential factors in buying garments. During the Focus Group 2 session, Makayla took the winter coat as an example and mentioned that price and quality were crucial as she seldom wore it and it should keep her warm. Kelly agreed and proposed that durability and materials used were essential as well:

Makayla: "For me, price and quality are the most important factors that I look for in choosing a garment, for example, when I need to buy a thick coat, I hope that it can be in high quality which can keep me warm and not so expensive as we actually only have to wear a thick coat for few days in a year in Hong Kong."

Kelly: "Yes, speaking about winter garments, I hope that I can wear them for several years so durability is what I care about. I would like to buy sweaters and coats with high durability. Materials used are also important. I like wool-blend sweaters with cashmere as they are warm and soft, but they are generally expensive. Therefore, durability and quality are both important to me."

Vincent, Tiffany, and Christy in Focus Group 1 also reflected similar rational considerations and took sportswear as an example:

Vincent: "So I usually have a plan in buying which brand. For example, I do lots of sports, and I always want to buy sportswear. But I compare different sports brands, like Adidas and Nike. I'd like to compare their quality, material used and price before I decide which brand to buy.

Tiffany: "For sportswear, they are important properties to look for."

Christy: "More than just the price, I agree that both quality and material used are important. I am quite picky, not only sportswear, any garment has to be in good quality and I feel comfortable when I wear it."

In addition, Focus Group 1 led the discussion about the size, colors, design, and cutting. These rational considerations can influence whether the garment is easy to match and fits on a person, as reported by Tiffany, can also be linked to the concept of self-fulfillment in the last section. Alex and Vincent suggested that these factors contribute to the first impression and play an important role in how others perceive the person:

Alex: "So the size and color are important considerations when choosing a garment, because a garment can affect others' perceptions towards a person."

Vincent: "Yes these two factors are the first impression given and the most eye-catching elements."

Tiffany: "So they are my considerations too as I care how others perceive the garment very much. I hope others think that it's beautiful too."

Tiffany: "Easy to match is definitely important as this can prevent me from buying more garments to combine into a perfect outfit. If I buy some tops with certain cuttings, I have to buy particular bottoms to match the top."

More than just the factors above, consumers usually purchase goods for their values rather than their functions, therefore, the product image and brand name are recognized as motives [35]. Hilary in Focus Group 2 certified this idea:

Hilary: "It's common for people to buy clothes in well-known and bigger brands so that we can obtain quality assurance."

In terms of sustainability considerations, the conversations in both groups began with the perception toward sustainability. Ecological sustainability, social sustainability, and economic sustainability are the three main aspects discussed:

Alex: "I think based on what we learnt in secondary school, there are ecological sustainability, social sustainability, and economic sustainability. To me, I believe the fashion industry itself is already unsustainable."

Vincent: "Yes, I also remember what we learnt in school. I agree that this industry is environmentally unsustainable, it's hard to avoid buying unsustainable garments because I think the whole production process can harm the environment."

Regarding the ethical buying part in the theoretical framework, [39] claim that consumers rarely consider sustainability factors when they are making a purchasing decision. It is also quite difficult to take ethical issues into consideration in daily life (*ibid.*). Vincent and Jane suggested that sustainability issues hardly be the factors that affect purchasing decisions:

Vincent: "When it comes to the product itself, it's quite hard to know whether it's sustainable or not, maybe because I'm not familiar with the materials. I think most people are only attracted by the price and the look."

Jane: "I value sustainability to some extent and notice how the garment industry negatively affects the environment, but the appearance and the price get a higher priority in my decision in buying clothes. For example, I really love a pair of denim jeans and they are on sale, at that moment I won't remember a large amount of water is used in the production."

On the other hand, both groups mentioned the ways to deal with unwanted clothes, either selling them or donating them to physical charities, these clothes could be bought online or in second-hand stores:

Kelly: "I also learn to go for quality over quantity in order to pay some effort to protect the environment. Moreover, I'd donate my unwanted clothes to charity in order to provide them an "after-life"."

Tiffany: "Yes, and I may find some "treasures" there. I remember that textile products account for around 4% of the global landfills, so it's important to train a sustainable consumption pattern."

Therefore, sustainability could be a reason that affects consumption behavior. All the factors revealed above will be further identified and discussed in the next chapter.

4.3   Competence

Competence is regarded as the third element, which is the consumers' knowledge and awareness toward sustainability of a garment and the whole industry. Sustainable fashion consumption was the main theme, with the responsibilities of being sustainable, the way to receive information about sustainability, and the sustainable measures taken by apparel companies. The discussion continued to answer the second research question in this section: *What are the perceptions of consumers regarding sustainability in the garment industry?*

Regarding whether the participants care about the sustainable measures taken by apparel companies, both groups take a few brands as examples to specify sustainable policies. The participants in Focus Group 2 discussed general policies, such as those about the waste issues and the whole supply chain:

Makayla: "Continuous work is required to implement measures in order to improve the life cycle of an apparel."

Jane: "Like how to use or re-use, repair and recycle, many retail brands now establish methods to make actions become sustainable."

Hilary: "Also from the production of raw materials, design, manufacturing, to marketing and sale, these stages receive less attention, but I believe there are still many measures tackling the sustainable issues in these processes."

On the other hand, Focus Group 1 illustrated some ethical issues of the social aspect, which were about welfare and working condition of the employees:

Tiffany: "Speaking about sustainability measures in garment factories, a fair working condition should be provided in terms of the social aspect."

Alex: "I agree, many of them bring competitive wages and compensations for the employees now."

Vincent: "At the same time, establishing high safety and health standards. I believe that sustainable measures can increase the status of the company."

Alex: "Definitely, these can let the public know that the company does something more than earning profits as it cares for the employees."

To a large extent in general, the respondents in both groups would like to receive information about the sustainability background of a garment or a garment brand, thus, they were willing to be aware of these issues. Christy and Tiffany indicated that sustainability was widely discussed, and they could easily get access to the information:

Christy: "I think sustainability is highly debated so it's really easy to receive this kind of information."

Tiffany: "Yes, no matter what I want to receive, it's an increasingly covered theme in the media worldwide."

Kelly and Makayla in Focus Group 2 led the conversations to transparency and the methods that they could learn the sustainability information of the garment:

Kelly: "If the apparel brands provide transparent information, I'm willing to receive such information as it's easy to get access to."

Makayla: "I believe most of the people, including me, want to know the background and sustainability level. However, if the brand is not sustainable enough, it won't provide so much transparent information about sustainability to the public."

Kelly: "Yes, but if the brand pays certain efforts in sustainability, advertisements are often established in the media."

Makayla: "I always see that on Facebook and Instagram. I'd like to know the information about sustainability as it's a positive image of the company."

In terms of the parties who should bear the responsibilities of being sustainable in the garment industry, the fashion enterprises and consumers were both specified in the focus groups. Christy and Vincent in Focus Group 1, and Jane and Kelly in Focus Group 2 proposed that the brands were the major party who took the power to promote sustainable actions to manufacturing factories and consumers:

Christy: "I think all people involved bear the responsibility. I, as a consumer, should think twice before consuming. I am also engaged in the production process in a fashion brand, which means that I should try my best in establishing the ideas in the sustainable projects."

Vincent: "I believe that the main responsibility of textile companies is to modify the production and marketing strategies towards a larger sustainability."

Jane: "There's no doubt that companies bear the responsibility as they can collaborate with factories and suppliers which can respect the environment and the employees."

Kelly: "And they can offer rewards to encourage customers to make sustainable actions, which provide incentives for them to re-wear and recycle unwanted garments."

During the discussion in Focus Group 2, Makayla and Hilary presented that the government accounted for a major responsibility to sustainability issues in the garment industry:

Makayla: "Exactly, I believe that about the social aspect like human rights matter and working conditions, the responsibility goes to the company, as well as the government."

Hilary: "Yes, the government should implement policies in building a fair working condition and protect the labor."

Makayla: "Also, subsidies can be provided by the government to the companies in implementing such measures."

In conclusion, this section mentions and explains how the participants perceived sustainability and also the level of awareness and knowledge toward the related issues.

5 Discussion

This chapter emphasizes the purpose of this research, which is to identify the current key factors that influence consumer behavior and attitudes in Hong Kong in terms of sustainability issues in the garment industry. Key objectives in this project relate to consumer awareness toward the aspects of sustainable development in garment purchasing decisions, and the two research questions are answered and discussed.

5.1 What Are the Factors that Affect Consumers in Purchasing Clothes?

This part answers the first research question. Various factors can influence consumers in purchasing garments from the discussions among Hong Kong participants. Similar to the literature from [2] and [19], both psychological and sociological considerations can be involved in consumer behavior when consumers are making a purchasing decision. The price, including discounts and a sale, is the first important factor. The quality, durability, the level of comfort, and materials used, are the rational factors which can affect the purchasing decision. In addition, the colors, the size, the design, and the cutting, are seen as the general appearance of a garment, which can also influence how consumers perceive the garment and their decisions to consume it. Others' opinions, such as friends and family members, with the mood and emotions of the buyer, are likely to influence people to purchase clothes. Moreover, the brand name and the current trend play a role in the decision as well. More than just the rational and emotional factors, outfits for different occasions and the level to match the whole outfit take part in affecting the process of buying garments.

Before the moderator brought sustainability up in the focus group discussions, only one of the participants proposed factors related to sustainability, which is about protecting the environment. This may be due to the lack of concern in sustainability issues, and they may not consider this factor during a purchasing decision until others talk about them. As suggested in the previous studies, an ethical and sustainable behavior is trained by developing a habit [39]. Only when a habit is developed in considering the sustainability factor, can the whole practice be regarded as a sustainable consumption. Moreover, [29] mention that consumers are more likely to pay attention to environmental policies implemented by brands before a purchasing decision if they have stronger attitudes toward these issues. However, there are no fundamental modifications in consumer behavior of sustainable garments, and this may be due to the complexity of sustainable fashion, with individual and social factors as well as barriers [33]. As a result, with a rise in sustainability concern and awareness, consumers are likely to have many other priorities and factors in choosing their garments.

5.2 What Are the Perceptions of Consumers Regarding Sustainability in the Garment Industry?

This section answers the second research question. During the sessions of the focus groups, the participants generally viewed the garment industry as unsustainable. [17] mention that the generation of these participants are more environmentally and socially conscious than other previous generations as they have been taught and educated to be more environmentally friendly. Hence, consumers are more likely to pay attention to environmental policies implemented by companies before a purchasing decision if they have stronger attitudes toward these issues [19]. The participants started to avoid purchasing similar garments that they already had, and this action of buying less unnecessary clothes can be regarded as sustainable. Also, they sold or donated unwanted garments online or to physical charities, and they considered purchasing garments in second-hand stores.

Consumers are more willing to pay more for products from companies that are environmentally responsible if they have more knowledge about these issues [29]. The participants in this project may lack awareness of sustainability in their purchasing decisions. However, they were able to list some sustainable measures taken by several apparel brands. More than just the environmental aspect, the employees' welfare and working conditions as part of the social aspect were discussed. To a large extent, the participants want to receive transparent information about sustainability. They discussed that they could easily get access to the information from the company and the product when these had been increasingly covered in the media worldwide, such as advertisements and promotions. [39] argue that brands have the responsibility in turning sustainable information available and transparent to consumers, whereas consumers have the responsibility in developing an ethical lifestyle. The participants agree that both the apparel brands and consumers bear the responsibility in the sustainability field within the garment industry, and they also proposed that the government had the responsibility to provide subsidies and implement laws to protect the workers. This suggests that consumers in the project have a certain level of concern toward sustainability.

6 Conclusions

In this research, the consumer attitudes toward sustainability in the garment industry were explored, with a consumer study in Hong Kong. The empirical data included two focus group discussions. The sessions consisted of the participants' attitudes and perceptions toward this topic and were based on their personal thoughts and opinions. The results and analysis were explained in relation to the aim and the two research questions.

In conclusion, there is a lot of factors that play a crucial role in consumers' purchasing decisions in the garment industry. Regarding the answer to the first research question, the factors are mainly classified into rational and emotional considerations. Sustainability factors are not the main considerations during a purchasing decision. This may be due to the complexity of sustainable fashion and the fact that consumers value individual factors more. In addition, there is a struggle between needs and wants, which can lead to the problem of over-consumption. The participants in the focus group discussions interpreted that the fashion industry was generally unsustainable, such as the manufacturing process and disposal of clothes. A rise in media coverage of sustainability issues within the garment industry, increases consumers' concern and awareness toward sustainability. Thus, consumers may start to value the related factors in choosing their garments. The focus groups also concluded that a number of stakeholder groups are seen as responsible for being sustainable as they viewed sustainability as important in this industry. To a large extent, the participants want to receive transparent information about sustainability.

6.1 Implications

This study investigated and analyzed consumer behavior as social practice in the garment industry. This can support the apparel companies to understand the consumer attitudes and opinions toward sustainability in this industry. In order to raise the awareness of Hong Kong consumers toward sustainable consumption, it is important for the garment brands to promote their sustainable actions. The sustainable measures taken by the enterprises are likely to positively influence consumers' purchasing decisions and the companies can gain better reputations.

This research explained how Hong Kong citizens viewed sustainability in a consumer study and how their consumer behavior in purchasing garments changed. The information in this project could help the "green" marketing of the brands. After they understand the expectations of consumers, they can be responsive to the consumer's needs and wants. For instance, social labels and eco-labels can be established during management decisions.

By adjusting both the goods and services, trust and value can be built, especially the effective measures that trigger the purchasing intention of consumers. More than just ethical consumer behavior, a sustainable future requires collaboration between different parties who want to achieve the same sustainability target and goal.

6.2   Suggestions for Future Research

The contribution of this research is to enhance the understanding of consumers' perceptions and attitudes toward sustainability in the garment industry. Due to the limited number of participants in data collection, there is a suggestion of conducting a quantitative research method in terms of this topic. The project can go deeper and further with a larger sample size, where more ideas can be collected with more interviewees. In addition, future studies can discuss sustainability in the garment industry from the perspectives of other parties, such as the apparel brands and the government. This study identified and analyzed the issues from the perspectives of consumers only, however, opinions and thoughts from other stakeholders are essential as well to ensure a sustainable future.

Appendix 1

Interview guide
Introduction of subject
My name is Isabella. We are here today due to my master thesis project in Sustainable Development at Uppsala University. The discussion today is going to be recorded and transcribed for research purposes only. This will not be shared with anyone else. Also, it is important that only one person speaks at a time so there won't be any problems hearing everyone's thoughts in the recording. Everyone's opinion is of great importance in this research study. There are no right or wrong answers, and we just want a discussion. This session is important to me to obtain a general and reliable understanding of the subjects discussed.

1.   (Materials)

Discuss the decision-making process in consumption in general, and then in terms of purchasing garments in detail.

- - Do you have any plans in buying and consuming?
- - What is/are motivation(s) for consumption?
- - What kind of properties are you looking for in the garment you purchase?

2. (Meanings)

Discuss the factors that affect you in buying clothes, as well as the perceptions and views regarding sustainability in the field of fashion.

- - What is important for you when you choose a garment that you want to buy?
- - What are the priorities and values of the factors and considerations?
- - How do you perceive sustainability?
- - Does the garment's sustainability level affect your purchasing decision?

3. (Competences)

Discuss knowledge and awareness regarding sustainable fashion consumption.

- - Do you care the sustainable measures taken by apparel brands? (Chemical/waste/ethical issues?)
- - To what extent would you like to receive information about the sustainability background of an apparel? In what ways?
- - Who do you think account for the largest amount of responsibilities of being sustainable?
- (Consumer/Companies/Others?)

Appendix 2

Template of GDPR form

Sveriges lantbruksuniversitet
Swedish University of Agricultural Sciences

SLU ID: sing0003

04/28/2028

Processing of personal data in independent projects

When you take part in the independent project 'Consumer attitudes towards sustainability in garment industry - A consumer study in Hong Kong', SLU will process your personal data. Consenting to this is voluntary, but if you do not consent to the processing of your personal data, the research cannot be conducted. The purpose of this form is to give you the information you need to decide whether or not to consent.

You can withdraw you consent at any time, and you do not have to justify this. SLU is responsible for the processing of your personal data. The SLU data protection officer can be contacted at dataskydd@slu.se or by phone, 018-67 20 90. Your contact for this project is: Ng Si Kei Isabella, isabella5959@yahoo.com.hk, +85267180273 / +460790528689.

We will collect the following data about you: Consumption behavior in purchasing garments and perspectives towards sustainability.

The purpose of processing of your personal data is for the SLU student to carry out their independent project using a scientifically correct method, thereby contributing to research within the field of consumer attitudes.

You will find more information on how SLU processes personal data and about your rights as a data subject at www.slu.se/personal-data.

☐ I consent to SLU processing my personal data in the way described in this document. This includes any sensitive personal data, if such data is provided.

_____ Hong Kong, 28 Apr 2020__

Signature   Place and date

_____

Name in block letters

Mailing address: Box 7060, 75007 Uppsala   Phone: +46 18 67 10 00 (switchboard)
                                                   Mobile: 070-3661419
VAT no: SE202100281701   cecilia.mark-herbert@slu.se
www.slu.se

References

1. Ajzen I (1991) The theory of planned behavior. Organ Behav Hum Decis Process 50(2):179–211
2. Askegaard S, Solomon M, Bamossy G (2006) Consumer behavior: A European perspective. Harlow: Financial Times/Prentice Hall
3. Bhaduri G, Ha-Brookshire J (2011) Do transparent business practices pay? Exploration of transparency and consumer purchase intention. Clothing and Textile Research Journal 29(2):135–149
4. Bhattacharya C, Luo X (2006) Corporate social responsibility, customer satisfaction, and market value. J Mark 70(4):1–18
5. Birtwistle G, Moore C (2007) Fashion clothing—Where does it all end up? International Journal of Retail & Distribution Management 35(3):210–216
6. Black S (2008) Eco-Chic: The fashion Paradox. Black Dog Publishing, London
7. Casey M, Krueger R (2015) Focus groups: A practical guide for applied research (5th ed.)
8. Devinney T, Belk R, Eckhardt G (2010) Why Don't consumers consume ethically? Journal of Consumer Behavior 9:426–436
9. Elkington J (1999) Cannibals with forks: the triple bottom line of 21st century business. Capstone, Oxford
10. Eriksson P, Kovalainen A (2008). Introducing qualitative methods: Qualitative methods In business research. https://doi.org/10.4135/9780857028044
11. Exchange C (2007) Idling engine: Hong Kong's environmental policy in a ten-year stall 1997–2007. Civic Exchange, Hong Kong
12. Flores J, Alonso C (1995) Using focus groups in educational research. Eval Rev 19(1):75–93
13. Gam H (2011) Are fashion-conscious consumers more likely to adopt eco-friendly clothing? Fashion Marketing and Management: An International Journal 15(2):178–193
14. Ghezzi A, Todeschini B, Cortimiglia M, Callegaro-de-Menezes D (2017) Innovative and sustainable business models in the fashion industry: Entrepreneurial drivers, opportunities, and challenges. Bus Horiz 60(6):759–770
15. Gordon R (2013) Unlocking the potential of upstream social marketing. Eur J Mark 47:1525–1547
16. Goss J, Leinbach T (1996) Focus groups as alternative research practice. *Area* 28(2), 115–123. Grehan, E., Shaw, D., Hassan, L., Shiu, E. & Thomson, J., 2005. An exploration of values in ethical consumer decision-making. Journal of Consumer Behaviour, 4(3)
17. Grehan E, Shaw D, Hassan L, Shiu E, & Thomson J (2005). An exploration of values in ethical consumer decision-making. Journal of Consumer Behaviour 4(3)
18. Han J, Seo Y, Ko E (2016) Staging luxury experiences for understanding sustainable fashion consumption: A balance theory application. J Bus Res 74:162–167
19. Hawkins D, Mothersbaugh D (2010) Consumer behavior: Building marketing strategy. McGraw-Hill Irwin, New York
20. Hiller C (2010) Internal and external barriers to eco-conscious apparel acquisition. Int J Consum Stud 34(3):279–286
21. Homan R (1991) Ethics in Social Research. Longman, London
22. Huberman A, Miles M. & Saldaña J (2013) Qualitative data analysis: A methods sourcebook, (3rd ed.) SAGE Publications Inc. (ISBN 1452257876)
23. Joergens C (2006) Ethical fashion: Myth or future trend? J Fash Mark Manag 10(3):360–371
24. Jönsson J, Wätthammar T, Mark-Herbert C (2013) Consumer perspectives on ethics in garment consumption—perceptions of purchases and disposal. In Röcklinsberg H, Sandin P (eds) The ethics of consumption: the citizen, the market and the law. Wageningen Academic Publishers, The Netherlands (pp 59–63). http://www.wageningenacademic.com/Default.asp?pageid=8&docid=16&artdetail=Eursafe2013&webgroupfilter=1&)
25. Kadolph S, Pasricha A (2009) Millennial generation and fashion education: A discussion on agents of change. International Journal of Fashion Design, Technology and Education 2(2–3):119–126

26. Kim Y, Park H (2016) An empirical test of the triple bottom line of customer-centric sustainability: The case of fast fashion. Fashion and Textiles 3(1):1–18
27. Kitzinger J (1994) The methodology of focus groups: The importance of interaction between research participants. Sociology of Health 16(1):103–121
28. Kong H, Ko E, Chae H, Mattila P (2016) Understanding fashion consumers' attitude and behavioral intention toward sustainable fashion products: Focus on sustainable knowledge sources and knowledge types. J Glob Fash Market 7(2):103–119
29. Kozar J, Hiller C (2013) Socially and environmentally responsible apparel consumption: Knowledge, attitudes and behaviours. Social Responsibility Journal 9(2):315–324
30. Larsson A, Buhr K, Mark-Herbert C (2012) Corporate responsibility in the garment industry towards shared value (Chapter 16). In Torres AL, Gardetti MA (eds). Sustainable Fashion and Textiles. Greenleaf Publishing Ltd (ISBN 978–1–906093–78–5) (pp 262–276). https://www.greenleafpublishing.com/sustainability-in-fashion-and-textiles
31. Lee K (2009) Gender differences in Hong Kong adolescent consumers' green purchasing behaviour. J Consum Mark 26(2):87–96
32. Lee H, Hill J (2012) Young generation Y consumers' perceptions of sustainability in the apparel industry. Journal of Fashion Marketing and Management: An International Journal 16(4):477–491
34. Lundblad L. & Davies I (2016) The values and motivations behind sustainable fashion consumption. Journal of Consumer Behaviour 15(2)
34. Malhotra N (2010) Marketing research—An applied orientation (6th ed.). New Jersey: Pearson
35. Markkula A., Eraranta K, Moisander J (2010) Construction of consumer choice in the market: Challenges for environmental policy. International Journal of Consumer Studies 34:73–79
36. Maslow A (1943) A theory of human motivation. Psychol Rev 50(4):370–396
37. Micklin P (2006) The Aral Sea crisis and its future: An assessment in 2006. Eurasian Geogr Econ 47(5):546–567
38. Moore R, McNeill L (2015) Sustainable fashion consumption and the fast fashion conundrum: Fashionable consumers and attitudes to sustainability in clothing choice. Int J Consum Stud 39(3):212–222
39. Neville B, Carrington M, Whitwell G (2014) Lost in translation: Exploring the ethical consumer intention-behavior gap. J Bus Res 67:2759–2767
40. Pantzar M, Shove E (2005) Consumers, producers and practices—Understanding the invention and reinvention of Nordic walking. J Consum Cult 5(1):43–64
41. Reckwitz A (2002) Toward a theory of social practices: A development in culturalist theorizing. Eur J Soc Theory 5(2):243–263
42. Riege A (2003) Validity and reliability tests in case study research: A literature review with "hands-on" applications for each research phase. J Cetacean Res Manag 6:75–86
43. Roxas B, Castaneda M, Marte R, Martinez C (2015) Explaining the environmentally-sustainable consumer behaviour: A social capital perspective. Social Responsibility Journal 11(4):658–676
44. Shen D, Richards J, Feng L (2013) Consumers' awareness of sustainable fashion. The Marketing Management Journal 23(2):134–147
45. Sirgy M (1982) Self-concept in consumer behavior: A critical review. Journal of Consumer Research 9:287–300
46. Valor C (2007) The Influence of information about labour abuses on consumer choice of clothes. J Mark Manag 23(7):675–695
47. Warde A (2005) Consumption and theories of practice. J of Cons Cult 5(2):131–153
48. Watson M, Shove E, & Pantzar M (2012) The dynamics of social practice—Everyday life and how it changes. SAGE Publications Ltd., London
49. Winter S, Lasch R (2016) Environmental and social criteria in supplier evaluation—Lessons from the fashion and apparel industry. J Clean Prod 139:175–190
50. Zukin S, Smith J (2004) Consumers and consumption. Ann Rev Sociol 30:173–197

Luxury Fashion Consumption in Collaborative Economy: A Conceptual Framework

Sheetal Jain

Abstract Lately, a drastic shift has been observed in demand for luxury goods from 'ownership' to 'usership'. Consumers are re-assessing their priorities, resulting in change in their attitude and behaviour towards luxury. Yet, very few studies have been conducted in this domain to gain an in-depth understanding about what drives and prevents consumers to consume luxury goods on a collaborative basis. This study is based upon extensive review of literature and tries to fill this gap by classifying various factors affecting luxury fashion consumption in collaborative economy into six broad categories, viz., individualistic/collectivistic culture, egoistic (self-oriented) values, altruistic (others' oriented) values, cost value, identifiable risks and personal norms. This study also explores the moderating role of identifiable risks and personal norms in consumer's intention to participate in collaborative luxury fashion consumption (CLFC). Going further, an integrated conceptual framework is proposed in this study to provide holistic view about the key elements of CLFC based upon dual theoretical framework of theory of planned behaviour (TPB) and Stern's value-belief-norm theory.

Keywords Collaborative consumption · Culture · Luxury fashion goods · Theory of planned behaviour · Value-belief-norm theory

1 Introduction

The concept of luxury has evolved radically over time. Traditionally, luxury was destiny of 'elite-few' but with the exponential growth and democratization of luxury goods market, it is no more reserved to 'members of the upper echelon of society but also accessible and available for the masses' [21, p. 425]. Today, consumers are relishing material comfort much more than their earlier generations, leading to increased preference for experiential consumption. A drastic shift has been observed in demand for luxury goods from 'ownership' to 'usership' [127]. Luxury has

S. Jain (✉)
Luxe Analytics, Delhi, India

© The Author(s), under exclusive license to Springer Nature Singapore Pte Ltd. 2022
S. S. Muthu (ed.), *Sustainable Approaches in Textiles and Fashion*,
Sustainable Textiles: Production, Processing, Manufacturing & Chemistry,
https://doi.org/10.1007/978-981-19-0874-3_3

progressed from just being a way to show-off to life experiences and personal enrichment [20]. Lately, consumers are re-assessing their priorities, resulting in change in their attitude and behaviour towards luxury [126, 48]. They are questioning the need to possess items when they can easily rent or share them [13, 56, 66].

The idea of collaborative consumption (CC) is not new. However, its acceptance in luxury industry is little surprising. Luxury refers to exclusivity and display of opulence [117] while collaborative consumption, by its very nature, makes things affordable and accessible to the consumers [9, 115]. Still, there is growing demand for collaborative consumption in luxury fashion segment, particularly, among millennials [90]. Immense potential for start-ups such as 'Rent the Runway', 'Walk in my closet', 'Albright' and 'Bag, borrow or steal' is recognized as lesser and lesser number of luxury shoppers want to bind themselves to a single product rather, they want to enjoy the flexibility to try new items more frequently [47 104]. As per research by Future Foundation, about one-fourth of the consumers in the age group of 15–24 and one-fifth in the age group of 25–34 are in favour of the idea of having temporary access to luxury items by way of collaborative consumption [126].

The rising trend towards collaborative luxury consumption is mainly due to the fading charm of flamboyant, materialistic possession and increasing concern for wasteful consumerism. Over a decade, luxury has become a lot less showy [24]. As per a report by Pricewaterhousecoopers [83, p. 14], today, only one in two consumers agree with the statement- 'owning things is a good way to show status in society'. Various researchers have reflected that there has been significant change in the values, beliefs, attitude and behaviour of consumers, thereby leading to shift towards collaborative consumption [63, 105]. However, so far, to the authors' best knowledge, no study has tried to analyse the impact of values on consumers' attitude, norms and intention to participate in collaborative consumption in context of luxury fashion industry. This study tries to bridge this gap. The main objective of this study is to develop an integrated conceptual framework based upon the extant review of literature that explains the key factors affecting CLFC using theory of planned behaviour (TPB) [1] and value-belief-norm (VBN) theory [106]. This study tries to elucidate the relationship between values (altruistic, biosphere and egoistic values) Stern et al. [106] and variables of TPB (attitude, subjective norm, perceived behavioural control and purchase intention) [1], thus offering a distinct theoretic outlook in collaborative luxury fashion context. This is one of the pioneer studies to explore the moderating role of identifiable risks and personal norms in consumer's intention to participate in CLFC.

This study will help companies to gain deeper understanding of what motivates consumers to participate in shared luxury consumption and enable them to develop innovative and more sustainable business models. It will also provide insights to the academicians about how values, attitude, norms and behaviour are correlated in context of collaborative consumption in luxury domain.

The chapter is structured as follows: next section presents literature review followed by conceptual framework. The subsequent section puts forward discussion and implications of the study. The last section talks about limitations and directions for future research.

2 Literature Review

2.1 Conceptualizing Luxury

The term luxury is derived from the Latin word 'luxus', which denotes 'soft or extravagant living, sumptuousness, opulence' [20, p. 17]. Although, the concept of luxury is hard to define [75] and highly subjective [52], Rodrigues et al. [91], but previous studies have defined luxury as dream [22], expensive, superior quality [53], waste [116], exclusive [84], timeless [48] and symbolic [44]. From the extant review of literature, the following two factors have emerged as the key reason behind the consumption of luxury goods, extrinsic factors (signal wealth) and intrinsic factors (self-indulgence) [4, 42].

2.2 Conceptualizing Collaborative Fashion Consumption (CFC)

Iran and Schrader [36] refer to collaborative fashion consumption (CFC) as a platform which enables consumers to have access to already existing apparel instead of buying new fashion items through either providing alternative opportunities to obtain individual ownership (gifting, swapping, or second-hand) or providing usage options for fashion items possessed by others (sharing, lending, renting or leasing). CC is considered as a new trend towards sustainable consumption [5, 65]. CC in fashion leads to less consumption of resources as compared to buying new clothes made from virgin materials [36, 119].

2.3 Luxury and Collaborative Fashion Consumption (CFC)

Lately, due to the advancement of communication technologies and captivating phenomenon of collaborative consumption, new business models have emerged in the luxury fashion industry such as luxury renting, luxury fashion libraries, etc. [12, 108]. Collaborative consumption not only makes luxury accessible for those who cannot afford it but it also provides value proposition for affluent users who are hesitant to repeat the same item again. It permits the participants to occasionally indulge in the luxury lifestyle without the need to pay the full price for the privilege [126]. For example, start-ups such as Rent the Runway provides users the access to full range of high-end labels for special occasions. Their business model helps people to rent luxury brands for a minuscule amount and enables them to accomplish their dream without burden of ownership. In addition, it also cuts superfluous spending on these goods by the people who can afford to buy it [16, 88]. These novel business models not only help in fulfilling consumer needs but also help in alleviating the ecological burden of luxury fashion industry [64].

2.4 Role of Values in Collaborative Luxury Fashion Consumption (CLFC)

Values are defined as 'desirable trans-situational goals', varying in importance, that serve as guiding principles in people's lives' [101, p. 21]. Values perform a significant role in developing attitude and intention towards a particular behaviour [2]. Although, understanding key factors behind consumer's intention to participate in collaborative luxury consumption is a complex process, a closer examination of basic human values that explain and influence the behaviour [109], can provide insights on reasons for adoption of collaborative (luxury fashion) consumption [19, 63].

Values are presumed to encourage eco-friendly behaviours [85]. Individual's values are found to be related with ethical behaviour [2]. However, a study by Barnes and Mattsson [8] posited that values, attitude and norms impede collaborative consumption (CC). Various studies have concluded that individuals with high self-transcendent or altruistic (others' directed) values have greater keenness to participate in sustainable behaviour [99] than individuals with greater self-enhancement or egoistic (self-directed) values [101, 107].

2.5 Factors Affecting Intention to Participate in CLFC

Based on extensive review of literature, various factors affecting intention to participate in CLFC could be broadly classified into six categories, viz., individualistic/ collectivistic culture, egoistic (self-oriented) values, altruistic (others' oriented) values, cost value, identifiable risk and personal norms.

2.5.1 Individualistic/Collectivistic Culture

Cultural values play a significant role in shaping an individual's lifestyle and behaviour [45, 114]. In individualistic culture, the self is seen as independent of others where as in collectivistic culture, the self is seen as interdependent with others [112]. Collectivism emphasizes on society and group goals [87]. On the other hand, individualism focuses on one's personal goals [112]. Differences in the beliefs of the individualists and collectivists help to determine the types of value they want to attain from goods and services [122]. Luxury goods consumption is triggered by cultural patterns [129].

Past studies have confirmed that culture influences sustainable consumer behaviour [100]. In western societies, consumption is a natural part of everyday life and over-consumption is widespread and therefore, resources are overexploited enormously. While, collectivistic culture is linked with sacrifice, interdependence and harmony [111]. Xu et al. [124] found that Chinese participants perceived high level of ecological value for consuming second-hand garments. In addition, for Chinese consumers most significant factor with respect to purchasing second-hand clothing was subjective norm while for American consumers, hedonic value was the most significant factor. Therefore, the following hypothesis is proposed:

P1: Individualistic culture and egoistic (self-oriented) values are positively related to each other with regard to consumer's intention to participate in CLFC.

P2: Collectivistic culture and altruistic (other'-oriented) values are positively related to each other with regard to consumer's intention to participate in CLFC.

2.5.2   Egoistic (Self-Oriented) Values

The customers with 'independent self-concept' are self-oriented. They stress on hedonic, materialistic and self-perception values [67]. Kucukemiroglu [58] in his study on different consumer segments with varied lifestyles revealed that self-directed pleasure is an important dimension in luxury goods consumption. Therefore, individuals who emphasize on egoistic (self-oriented) values are less in favour of CC as they may find it difficult to forgo individual ownership [37]. A study by Ross and Haln (2017) concluded that consumers who participated in CC did not care for themselves rather they were considerate for others. Iran and Geiger [37] revealed that egoistic value had negative relationship with attitude towards collaborative fashion consumption whereas biosphere and altruistic values were positively related. Therefore, the following hypothesis is proposed:

P3: Egoistic (self-oriented) values and attitude are negatively related to each other with regard to consumer's intention to participate in CLFC.

Based on the extensive review of literature, this study posits the following egoistic (self-oriented) values as key factors affecting luxury fashion consumption on a collaborative basis:

- Hedonic Value
- Materialistic Value
- Achievement Orientation
- Variety Orientation

Hedonic value: Hedonic value refers to consuming luxury goods mainly for self-indulgence rather than utilitarian benefits. Hedonistic shoppers are keen in expressing themselves and are ready to pay extra for luxury goods [118]. Therefore, they might deter from adopting collaborative consumption models. However, few people may perceive pleasure in sharing concepts like swapping parties [37]. Young consumers who cannot afford to buy expensive labels, enjoy immediate gratification through temporary access of luxury goods [70, 74]. Roux and Guiot [97] and Hamari et al. [31] posited hedonic value as key driver behind collaborative consumption or sharing-related activities. CC gives opportunity to participants to look for unique products at reasonable prices and is also linked with thrilling experience of searching something of real value at affordable prices [71, 120]. Therefore, the following hypothesis is proposed:

P4: Hedonic value and attitude are positively related to each other with regard to consumer's intention to participate in CLFC.

Materialistic value: Materialistic value refers to the value attached to the material objects by a person [10]. Materialistic individuals constitute major proportion of luxury consumer segment [121]. Many studies have revealed positive relationship between materialistic value and luxury goods consumption [117, 122]. Consuming a luxury product enhances the self-concept of the user. These consumers consider their belongings as part of their identity [118]. Individuals who attach greater importance to ownership may not be willing to engage in CC [37, 71]. Schwartz and Boehnke [102] argued that materialism and green behaviour are antithetical to each other. Several studies have found positive relationship between materialism and self-enhancement (egoistic values) [30, 28] whereas negative relationship between materialism and self-transcendent (altruistic values) [55]. Tilikidou and Delistavrou [110] confirmed that likelihood of over-consumption is more when an individual has materialistic values. Therefore, materialistic orientation inherently counters the basic idea of CLFC. Hence, the following hypothesis is proposed:

P5: Materialistic value and attitude are negatively related to each other with regard to consumer's intention to participate in CLFC.

Achievement Orientation: Consumers use luxury as status symbol to demonstrate their professional success and achievement to others [73, 78]. Jain [40] found that men feel sense of achievement and pride whereas women feel delighted and pampered through their luxury goods consumption. Hamari et al. [31] argued that active participation in CC may provide sense of achievement to the individual within CC community. Collaborative luxury consumption enables consumers to get access to high-end brands which otherwise they may not afford and exhibit their success and achievement [47]. Therefore, the following hypothesis is proposed:

P6: Achievement orientation and attitude are positively related to each other with regard to consumer's intention to participate in CLFC.

Variety orientation: Intrinsic aspirations and self-directed pleasure are found to be significantly related with luxury brand consumption [113]. Collaborative luxury consumption provides consumers the pleasure of shopping [61] through the satisfaction of their need for variety and variation [128] without feeling of guilt [54, 64]. Moeller and Wittkowski [71] stated a positive link between fashion (variety) orientation and choice of non-ownership mode of consumption. New concepts like fashion libraries provide members an opportunity to experiment with styles and looks without having to pay full price. Pedersen and Netter [81] claimed that variety and style may be more important to engage consumers in collaborative fashion consumption than making sustainability appeal to consumers. Therefore, variety orientation is assumed to have positive impact on CLFC. Hence, the following hypothesis is proposed:

P7: Variety orientation and attitude are positively related to each other with regard to consumer's intention to participate in CLFC.

2.5.3 Altruistic (others'-Oriented) Values

The more an individual's values show concern for others and environment, the greater will be the expectations of others to act sustainably (Roos and Hahn 2017). Prior studies confirmed that sharing consumers have greater altruistic and biosphere values than non-sharing consumers [85]. Altruistic values are assumed to have positive relationship with norms, resulting in eco-friendly behaviour [106]. Therefore, the following hypothesis is proposed:

P8: Altruistic (others'-oriented) values and subjective norm are positively related to each other with regard to consumer's intention to participate in CLFC.

Based on the extensive review of literature, this study posits following altruistic (others-oriented) values as key factors affecting luxury fashion consumption on a collaborative basis:

- Social Value
- Symbolic Value
- Ecological Value

Social value: Consumption of luxury brand enhances the social status of the user [117]. Botsman and Rogers [13] posited that individuals are indulging in new social norms that support collaborative consumption rather than individual consumption. Individuals develop and convey a socially conscious self through re-using clothes [17], as it lessens new garment production [34], thereby resulting in reduced use of resources and elongation of product life cycle [51, 64]. Worldwide, there is a growing trend towards socially and environmentally conscious consumption behaviour [45]. Collaborative fashion consumption contributes to the trend of being socially responsible [62]. Drive for voluntary simplicity associated with societal and environmental concerns is an upcoming trend [123].

Earlier, sharing was common among friends, family, relatives, etc., however, technology has enabled to scale it like never before [37]. Gaining social value has been found to be a key factor in determining participation in communities and online collaboration activities [79]. Participants engage in social interactions during swapping parties and develop long-term relationships with each other. Therefore, the following hypothesis is proposed:

P9: Social value and subjective norm are positively related to each other with regard to consumer's intention to participate in CLFC.

Symbolic value: High symbolic value and psychological benefits are the key factors driving luxury purchase decisions [121]. Belk [11] emphasized on self-concept and focused on the importance of acquiring luxury brands as it exhibits owner's identity. Individuals who buy or aspire to buy high-end labels mainly to symbolize status may explore the option of consuming these goods on a collaborative basis [23]. The growing popularity of collaborative consumption models like short-term rentals, has led to a situation, where it is hard for significant others to state if the user of an item is the actual owner [12], thereby giving an opportunity to the user to flaunt and express himself/herself. Leifhold [63] argued that such symbolic product consumption may help consumers to exhibit certain status at affordable prices and enhance their social acceptance and self-esteem [32, 41, 76]. It provides them a chance to consume luxury items as a symbolic marker of group orientation [98, 103]. Therefore, the following hypothesis is proposed:

P10: Symbolic values and subjective norms are positively related to each other with regard to consumer's intention to participate in CLFC.

Ecological value: Concern for planet drives consumer's readiness to participate in alternative consumption models [6, 125]. Young consumers have exhibited strong interest to support environmental sustainability [69]. Hamari et al. [31] argued that sustainable consumption is the main driver behind consumer's intention for CC. Sustainable behaviour augments an individual's status in the society, as per Griskevicius et al. [29]: 'going green to be seen'. Therefore, individuals with high ecological value are assumed to have greater intent to engage in CC of luxury goods. Hence, the following hypothesis is proposed:

P11: Ecological value and subjective norm are positively related to each other with regard to consumer's intention to participate in CLFC.

2.5.4 Cost Value

Luxury fashion brands are characterized by exclusivity, premium prices, image and status [39]. But not all the consumers who want access to these status brands are prepared to spend what these brands cost [82]. Monetary savings are presumed to be a key incentive of CC [63, 72]. Previous studies (Roos and Hahn 2017) revealed the importance of economic benefits such as cost-saving and utility as the main reason for CC. CLFC provides faster and economical way to fulfil consumers' dreams [60]. Pedersen and Netter [81] argued young consumers who are concerned about their appearance, have restricted monetary resources and believe in sustainable practices are open to alternative consumption models like fashion libraries. Key factor behind growing popularity of CE is that consumers, particularly millennials, are losing interest in all the costs and troubles associated with owning and maintaining a good. Therefore, they are opting for varied alternative solutions like renting, swapping, etc. [59]. Hence, the following hypothesis is proposed:

P12: Cost value and perceived behavioural control are positively related to each other with regard to consumer's intention to participate in CLFC.

2.5.5   Identifiable Risk as a Moderator

CLFC is a nascent concept. Various risks are perceived with collaborative luxury fashion consumption such as hygiene issues, quality aspects, non-existence of faith and absence of title [49]. Some people find wearing pre-owned clothes as embarrassing [37] and there is a stigma attached to it which prevents from adoption of CFC [27]. Armstrong et al. [7] pointed out consumers' consideration regarding bugs, cleanliness of garments and transmission of bacteria, diseases, odour and dirtiness from previous owners [83, 95]. Catulli [15] claimed that lack of ownership through sharing or renting models may adversely influence the need for self-expression. Various studies in the past have found negative relationship between perceived risk and intention to participate in CC [115]. Therefore, the following hypothesis is proposed:

P13: Identifiable risk moderates the effect of (a) Attitude (b) Subjective norm and (c) Perceived behavioural control on consumer's intention to participate in CLFC, such that TPB variables will have a more positive impact on consumer's intention to participate in CLFC, when identifiable risk is lower than higher.

2.5.6   Personal Norms as a Moderator

Personal norms refer to an individual's own moral obligation or responsibility regarding performance or non-performance of a behaviour, beyond perceived social pressure [1]. A sense of responsibility towards society and environment is growing among opulent members of the society [46]. They are looking for sustainable options [57] that reflect their own values and beliefs [33]. The greater the personal norm to consume luxury goods on a collaborative basis, the stronger will be the intention to do so [93]. Therefore, the following hypothesis is proposed:

P14: Personal norms moderates the effect of (a) Attitude (b) Subjective norm and (c) Perceived behavioural control on consumer's intention to participate in CLFC, such that TPB variables will have a more positive impact on consumer's intention to participate in CLFC, when personal norms are stronger than weaker.

2.6   Theory of Planned Behaviour

Theory of planned behaviour (TPB) [1] is used as a theoretical basis for this study as this theory plays a key role in explaining both self-oriented as well as others'-oriented values when used in behavioural studies [43, 77]. This theory is an addition to theory of reasoned action (TRA) and is established to overcome the limitations in the TRA [2, 26]. TPB posits that attitude, subjective norm and perceived behavioural control (PBC) are antecedents to consumers' behavioural intention and behaviour. Leifhold

[63] concluded that attitude, subjective norm and PBC have significant impact on consumers' intention to engage in CFC.

Table 1 presents review of studies performed in context of collaborative or shared consumption using TPB [31], Ross and Hahn (2017) [63, 62]. But, based upon extensive review of literature, no study has been performed in context of luxury. To bridge this literature gap, this study is conducted.

3 Conceptual Framework

The theoretical basis of this study rest upon dual conceptual framework of the TPB and VBN theory. This framework helps in explaining the relationship between values (egoistic and altruistic values) and TPB variables in context of CLFC. The extension of TPB helps to better explain consumer behaviour [1]. Based upon extant review of literature, the conceptual framework for this study depicts various factors affecting CLFC (Fig. 1). It exhibits that intention to participate in CC is influenced by an individual's attitude, subjective norm, PBC, identifiable risks and personal norms. Further, attitude is determined by egoistic (self-oriented) values, subjective norm is determined by altruistic (others'-oriented) values and PBC is determined by the cost value. In turn, egoistic and altruistic values are determined by individualistic/collectivistic culture. This study is based upon the prior research work [18, 25], to associate values, attitudes and norms in behavioural theories to better predict participants' collaborative consumption behaviour.

4 Discussion and Implications

The main objective of this study is to understand what motivates consumers to engage in collaborative luxury consumption. The theoretical foundation of this study is based upon value-belief-norm (VBN) theory [106] and TPB [1]. This is one of the first studies that provide holistic view about factors affecting luxury goods consumption on a collaborative basis. Yeoman [126] mentioned the growing popularity of alternative luxury consumption models like renting. However, so far, no study, to the authors' best knowledge have tried to develop an integrated conceptual model that explains key reasons behind collaborative consumption intention among consumers in luxury fashion context using well-established consumer behaviour theories.

This study has several important practical and theoretical implications. Firstly, it provides thorough understanding about various factors that drive and prevent consumers' participation in CLFC. This will help marketers to devise suitable strategies to encourage greater participation among consumers in this new economy. Some consumers participate in alternative consumption models for egoistic/self-oriented values like hedonic value while rest may participate for altruistic/others'-oriented values like social, symbolic and ecological value. Therefore, marketers must devise

Table 1 Review of studies performed in context of collaborative or shared consumption using theory of planned behaviour

Author(s)	Focus	Key Findings	Country	Type of study
Brandão and Costa [14]	To measure the relative importance of different barriers to sustainable fashion consumption	Findings confirm the role of TPB cognitions on predicting intention and show that the proposed barriers provide a satisfactory explanation of the TPB model	Europe, Asia, North America	Quantitative
Leifhold [63]	To understand the influence of consumer's values, attitude and norms on intention to participate in CC	Attitude, subjective norm and PBC were found to have significant impact on consumers' intention to engage in CC	Germany	Quantitative
Iran and Geiger [37]	To understand the impact of biospheric, altruistic, hedonic and egoistic values on attitude towards CFC	Egoistic values are found to have negative impact on attitude towards CFC while biosphric value has positive impact	Germany	Combination of Qualitative and Quantitative
Iran et al. [38]	This study employed TPB to test for the case of collaborative fashion consumption in cross-cultural context	Attitude, subjective norm and PBC were found to be relevant predictors of intention to adopt CFC	Tehran & Berlin	Quantitative
Roos and Hahn (2017)	This study used TPB and VBN theory to understand the effect of shared consumption on consumers' values, attitudes and norms	Shared consumption has significant positive effect on alruistic values, attitudes, subjective norms and personal norms	Germany	Quantitative

(continued)

Table 1 (continued)

Author(s)	Focus	Key Findings	Country	Type of study
Roos and Hahn [94]	The study employed extended TPB to understand the impact of consumer's personal norms, attitudes, subjective norms and PBC on CC	CC is more strongly influenced by personal norms and attitude than subjective norms. Personal norms with regard to CC is determined by altruistic and biospheric values orientations	Germany	Combination of Qualitative and Quantitative
Lindblom et al. [66]	To understand how materialism and price consciousness are related to attitude towards CC and their intention to engage in such behaviour	Materialism is found to be negatively related to consumers' attitude towards CC. However, materialism is found to be positively related to consumers' intention to engage in CC. Price consciousness is found to be positively related to consumers' attitude and intention to engage in CC	Finland	Quantitative
Lang and Armstrong [62]	To examine the influence of personality traits on consumers' intention to engage in CC	The findings revealed that fashion leadership, need for uniqueness and materialism influences attitude, subjective norm and PBC, thereby, influencing consumers' intention to rent and swap clothing	US	Quantitative

(continued)

Table 1 (continued)

Author(s)	Focus	Key Findings	Country	Type of study
Xu et al. [124]	To examine young consumers' behaviour towards second-hand clothing	The American participants had more positive response with regard to owning second-hand clothing items than Chinese consumers	US + China	Quantitative
Hwang and Griffiths [35]	To examine how cognitive value perceptions and affective attitudes are related to behavioural intention in context of CC	Results confirmed that various value perceptions (utilitarian, hedonic and symbolic) had varied influence on consumers' attitude and empathy towards CC services	US	Quantitative
Hamari et al. [31]	To understand why people participate in CC	Results showed intrinsic motivations (enjoyment and sustainability) and extrinsic motivations (economic benefits and reputation) as key factors behind participation in CC	Finland	Quantitative

Source Author's analysis

suitable strategies for each segment. For instance, marketers should highlight on augmenting consumers' social standing and prestige through consumption of luxury brands on a collaborative basis. Advertising message such as 'why wear the same, when each time you can try new' can be used to encourage consumers (others'-oriented) to switch to collaborative platforms. Likewise, for self-oriented consumers, luxury rental companies should offer wide variety of brands, designs and models, ranging from classics to the latest editions. RenttheRunway, use advertising tag lines 'Unlock an endless wardrobe' to entice variety-seeking shoppers [80].

Companies should also work towards removing various risks perceived by consumers in using luxury goods on a collaborative basis. This could be easily served by providing consumers, standard quality, clear value preposition, evidence

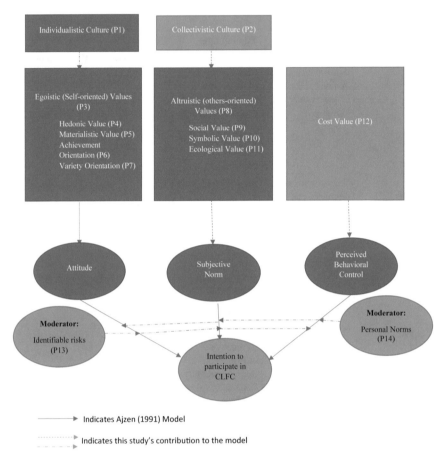

Fig. 1 Conceptual framework based on extended theory of planned behaviour. *Source* Author's analysis

of certified cleaning, guarantee, insurance, etc. [125]. Collaborative business models should be executed by well-known companies to develop trust among customers [6]. For example, in 2018, iconic Swiss watch brand Audemars Piguet declared the launch of its pre-owned luxury watch store [89]. Such a move will create confidence among consumers regarding product genuineness and encourage them to build positive attitude towards pre-loved/rental products.

In addition, companies can educate consumers about contribution of CFC in sustainability as transition towards ownerless models requires much more comprehensive change in consumers' mindsets [83]. Marketers should stimulate sense of ownership among shoppers through creative marketing campaigns such as 'access is the new ownership' or 'why buy when you can rent' and inspire shoppers to experiment new forms of consumption. Further, marketers should enlighten shoppers about how collaborative business models lessen the use of virgin materials, prolong product use time, eliminate waste and help industry to close material loops.

Secondly, this study will help marketers to examine how different cultures put influence on consumers' values and motivation for CC behaviour. Collectivism promotes consideration for people and planet and is governed by altruistic and ecological values [68] while individualism supports consideration for self and is mainly guided by egoistic values [3]. Therefore, marketers need to develop country-specific strategies based on the cultural background to encourage collaborative practices among consumers.

Lastly, this research will add to the existing body of knowledge about various factors affecting collaborative consumption intention in luxury fashion context. To the authors' best knowledge, so far, no study has provided a comprehensive framework on the key motivators behind participation in CC for luxury fashion goods using well-established consumer behaviour theories. This study also explores the moderating role of identifiable risks and personal norms in consumer's intention to participate in CLFC. This study will provide deeper insights to the academicians about what drives and prevents consumption of luxury goods on a collaborative basis.

5 Limitations and Directions for Future Research

This study has few limitations that pave the way forward towards future research work. This study focused upon consumers' perspective regarding participation on a collaborative basis for luxury fashion goods. However, in future, studies can be conducted to understand organizational drivers and barriers for CC in luxury domain. This study limits itself to luxury fashion segment, therefore, future studies may explore other luxury segments like automobiles, home, hospitality, etc., with regard to CC. This research is exploratory in nature. The conceptual model developed in this study may be further validated by future studies through empirically testing the proposed hypothesis.

References

1. Ajzen I (1991) The theory of planned behaviour. Organ Behav Hum Decis Process 50:179–211
2. Ajzen I, Fishbein M (1980) Understanding attitudes and predicting social behaviour. Prentice-Hall, Englewood Cliffs, NJ
3. Amatulli C, Angelis MD, Costabile M, Guido G (2017) Sustainable luxury brands: Evidence from research and implications for managers. Palgrave Macmillan, UK
4. Amatulli C, Guido G (2011) Determinants of purchasing intention for fashion luxury goods in the Italian market - A laddering approach. J Fash Mark Manag 15(1):123–136
5. Armstrong CM, Lang C (2013) Sustainable product service systems: the new frontier in apparel retailing? Res J Text Appar 17(1):1–12

6. Armstrong CM, Niinimäki K, Kujala S, Karell E, Lang C (2015) Sustainable product-service systems for clothing: exploring consumer perceptions of consumption alternatives in Finland. J Clean Prod 97:30–39

7. Armstrong CM, Niinimäki K, Lang C, Kujala S (2016) A use-oriented clothing economy? Preliminary affirmation for sustainable clothing consumption alternatives. Sustain Dev 24(1):18–31

8. Barnes, S. J., & Mattsson, J. (2016), "Understanding current and future issues in collaborative consumption: A four-stage Delphi study", Technological Forecasting & Social Change. Retrieved from https://doi.org/10.1016/j.techfore.2016.01.006

9. Battle A, Ryding D, Henninger CE (2018) Access-based consumption: a new business model for luxury and secondhand fashion business? In: Ryding D, Henninger C, Blazquez Cano M (eds) Vintage luxury fashion. Palgrave advances in luxury. Palgrave Macmillan, Cham

10. Belk RW (1985) Materialism: trait aspects of living in the material world. J Consum Res 12(3):265–280

11. Belk RW (1988) Possessions and the extended self. J Consum Res 15(2):139–168

12. Belk R (2014) You are what you can access: sharing and collaborative consumption online. J Bus Res 67(8):1595–1600

13. Botsman R, Rogers R (2011) What's mine is yours: the rise of collaborative consumption. Harper Business, New York City.

14. Brandão A, da Costa AG (2021) Extending the theory of planned behaviour to understand the effects of barriers towards sustainable fashion consumption. Eur Bus Rev 33(5):742–774. https://doi.org/10.1108/EBR-11-2020-0306

15. Catulli M (2012) What uncertainty? J Manuf Technol Manag 23(6):780–793

16. Cauthen CE (2014) 5 Websites to rent clothes from right now. Available at http://thestylishstandout.com/2014/04/16/where-to-rent-clothes-from/. Accessed 23 Sept 2018

17. Cervellon MC, Carey L, Harms T (2012) Something old, something used: determinants of women's purchase of vintage fashion vs. second-hand fashion. Int J Retail Distrib Manag 40:956–974

18. Conner M, Armitage CJ (1998) Extending the theory of planned behaviour: a review and avenues for further research. J Appl Soc Psychol 28(15):1429–1464

19. Connolly J, Shaw DS (2006) Identifying fair trade in consumption choice. J Strateg Mark 14(4):353–368

20. Danziger PN (2005) Let them eat cake: marketing luxury to the masses – as well as the classes. Dearborn Trading Publishing, New York, USA

21. Doss F, Robinson T (2013) Luxury perceptions: luxury brand vs. counterfeit for young USA female consumers. J Fash Mark Manag 17(4):424–439

22. Dubois B, Czellar S (2002) Prestige brands or luxury brands? an exploratory inquiry on consumer perceptions. Marketing in a Changing World: Scope, Opportunities and Challenges: Proceedings of the 31st EMAC Conference, University of Minho, Portugal, 28–31 May.

23. Durgee JF, Colarelli O'Connor G (1995) An exploration into renting as consumption behaviour. Psychol Mark 12:89–104

24. Eastman JK, Iyer R, Shepherd CD, Heugel A, Faulk D (2018) Do they shop to stand out or fit in? The luxury fashion purchase intentions of young adults. Psychol Mark 35(3):220–236

25. Ferrell OC, Crittenden VL, Ferrell L, Crittenden WF (2013) Theoretical development in ethical marketing decision making. AMS Review 3(2):51–60

26. Fishbein M, Ajzen I (1975) Belief, attitude, intention, and behaviour: an introduction to theory and research. Addison-Wesley, Reading, MA

27. Fisher T, Cooper T, Woodward S, Hiller A, Goworek H (2008) Public understanding of sustainable clothing: a report to the Department for Environment, Food and Rural Affairs. London, UK. Available at http://randd.defra.gov.uk/Document.aspx?Document=EV0405_7666_FRP.pdf. Accessed 21 Oct 2017

28. Gatersleben B, White E, Abrahamse W, Jackson T, Uzzell D (2010) Values and sustainable lifestyles. Archit Sci Rev 8628(53):37–50

29. Griskevicius V, Tybur JM, Van den Bergh B (1943) Going green to be seen: status, reputation and conspicuous conservation. J Pers Soc Psychol 98(2548):392–404
30. De Groot JIM, Steg L (2008) Value orientations to explain beliefs related to environmental significant behaviour: how to measure egoistic, altruistic, and biospheric orientations. Environ Dev 40(3):330–354
31. Hamari J, Sjöklint M, Ukkonen A (2015) The sharing economy: Why people participate in collaborative consumption. J Am Soc Inf Sci 67(9):2047–2059
32. Hann Y (2011) Cross-cultural understandings of 'face' and their influences on luxury brand consumption: a comparison of British and Korean attitudes and practices. J Glob Fash Market 2(1):36–43
33. Hennigs N, Wiedmann KP, Klarmann C, Behrens S (2013) yev as part of the luxury essence: delivering value through social and environmental excellence. J Corp Citizsh 52:25–35
34. Hu ZH, Li Q, Chen XJ, Wang YF (2014) Sustainable rent-based closed-loop supply chain for fashion products. Sustainability 6(10):7063–7088
35. Hwang J, Griffiths MA (2017) Share more, drive less: millennials value perception and behavioral intent in using collaborative consumption services. J Consum Mark 34(2):132–146. https://doi.org/10.1108/JCM-10-2015-1560
36. Iran S, Schrader U (2017) Collaborative fashion consumption and its environmental effects. J Fash Mark Manag Int J 21(4):468–482
37. Iran S, Geiger SM (2018) Eco friendly and fair: fast fashion and consumer behavior (eds: Heuer M, Becker-Leifhold C). Taylor & Francis, London, pp 153–162
38. Iran S, Geiger SM, Schrader U (2018) Collaborative fashion consumption- A cross cultural study between Tehran and Berlin. J Clean Prod 212:313–323
39. Jackson T (2004) A comparative analysis of global luxury brands. International retail marketing. Elsevier Butterworth-Heinemann, Oxford
40. Jain S (2019) Exploring relationship between value perception and luxury purchase intention: a case of Indian millennials. J Fash Mark Manag 23(4):414–439
41. Jain S (2021) Role of conspicuous value in luxury purchase intention. Mark Intell Plan 39(2):169–185
42. Jain S, Khan MN, Mishra S (2015) Factors affecting luxury purchase intention: a conceptual framework based on an extension of the theory of planned behaviour. South Asian J Manag 22(4):136–163
43. Jain S, Khan MN, Mishra S (2017) Understanding consumer behaviour regarding luxury fashion goods in India based on the theory of planned behaviour. J Asia Bus Stud 11(1):4–21
44. Jain S, Shankar A (2021) Exploring gen Y luxury consumers' webrooming behavior: an integrated approach. Australas Mark J. https://doi.org/10.1177/2F18393349211022046
45. Jain S (2018) Factors affecting sustainable luxury purchase behaviour: a conceptual framework. J Int Consu Mark 31:130–146
46. Jain S, Mishra S (2019) Sadhu—on the pathway of luxury sustainable circular value model. In: Gardetti M, Muthu S (eds) Sustainable luxury. Environmental footprints and eco-design of products and processes. Springer, Singapore, pp 55–82
47. Jain S, Mishra S (2020) Luxury fashion consumption in sharing economy: a study of Indian millennials. J Glob Fash Mark 11:171–189
48. Jain S (2020) Doodlage: reinventing fashion via sustainable design. In: Muthu S, Gardetti M (eds) Sustainability in the textile and apparel industries. Sustainable textiles: production, processing, manufacturing & chemistry. Springer, Cham
49. Joung H-M (2013) Materialism and clothing post-purchase behaviours. J Consum Mark 30(6):530–537
50. Kaiser FG (1990) The social psychology of clothing: symbolic appearances in context, 2nd edn. Machmillan, New York
51. Kale GO, Öztürk G (2016) The importance of sustainability in luxury brand management. Inter Int e-J 1(3):106
52. Kapferer J-N (1997) Managing luxury brands. J Brand Manag 4(4):251–260

53. Kapferer JN (2001) Reinventing the brand: can top brands survive the new market realities? Kogan Page, London
54. Kendall J (2010) Responsible luxury: a report on the new opportunities for business to make a difference. Available at https://www.rapaportfairtrade.com/Docs/CIBJO-responsible_luxury. pdf. Accessed 23 Sept 2018
55. Kilbourne W, Grünhagen M, Foley J (2005) A cross-cultural examination of relationship between materialism and individual values. J Econ Psychol 26:624–641
56. Kim NL, Jin BE (2019) Why buy new when one can share? Exploring collaborative consumption motivations for consumer goods. Int J Consum Stud 44(2):122–130
57. Kong HM, Eunju K, Chae H, Mattila P (2016) Understanding fashion consumers' attitude and behavioural intention toward sustainable fashion products: focus on sustainable knowledge sources and knowledge types. J Glob Fash Mark 7:103–119
58. Kucukemiroglu O (1999) Market segmentation by using consumer lifestyle dimensions and ethnocentrism: an empirical study. Eur J Mark 33(5/6):470–487
59. Kymalainen A (2015) Exploring motivations to engage in collaborative consumption. Available at http://epub.lib.aalto.fi/en/ethesis/pdf/14158/hse_ethesis_14158.pdf. Accessed 23 Sept 2018
60. Kärkkäinen H (2013) Renting luxuries as an identity project - a hermeneutic approach. Available at http://epub.lib.aalto.fi/en/ethesis/pdf/13285/hse_ethesis_13285.pdf. Accessed 23 Sept 2018
61. Laitala K (2014) Consumers' clothing disposal behaviour - a synthesis of research results. Int J Consum Stud 38(5):444–457
62. Lang C, Armstrong CMJ (2018) Collaborative consumption: the influence of fashion leadership, need for uniqueness, and materialism on female consumers' adoption of clothing renting and swapping. Sustain Prod Consum 13:37–47
63. Leifhold C (2018) The role of values in collaborative fashion consumption-A critical investigation through the lenses of the theory of planned behaviour. J Clean Prod 199:781–791
64. Leifhold C, Iran S (2018) Collaborative fashion consumption- drivers, barriers and future pathways. J Fash Mark Manag Int J 22(2): 189–208
65. Liedtke C, Baedeker C, Hasselkuß M, Rohn H, Grinewitschus V (2015) User-integrated innovation in sustainable LivingLabs: an experimental infrastructure for researching and developing sustainable product service. J Clean Prod 97:106–116
66. Lindblom A, Lindblom T, Wechtler H (2018) Collaborative consumption as C2C trading: analyzing the effects of materialism and price consciousness. J Retail Consum Serv 44:244–252
67. Markus HR, Kitayama S (1991) Culture and the self: implications for cognition, emotion, and motivation. Psychol Rev 98(2):224–253
68. Milfont TL, Duckitt J (2006) Preservation and utilization: understanding the structure of environmental attitudes. Medio Ambiente y Comportamiento Humano 7:29–50
69. Mishra S, Jain S, Malhotra G (2021) The anatomy of circular economy transition in the fashion industry. Soc Responsib J 17(4):524–542
70. Mishra S, Jain S, Jham V (2020) Luxury rental purchase intention among millennials- a cross nation study. Thunderbird Int Bus Rev 63:503–516
71. Moeller S, Wittkowski K (2010) The burden of ownership: reasons for preferring renting. Manag Serv Qual 10(2):176–191
72. Möhlmann M (2015) Collaborative consumption: determinants of satisfaction and the likelihood of using a sharing economy option again. J Consum Behav 14(3):193–207
73. Neruekar S (2014) Meet the Indian Yummy. Sunday Times of India, New Delhi, May 25
74. Noble SM, Haytko DL, Phillips J (2009) What drives college-age Generation Y consumers? J Bus Res 62:617–628
75. Nueno JL, Quelch JA (1998) The mass marketing of luxury. Bus Horiz 41(6):61–68
76. O'Cass A, Frost H (2002) Status brands: examining the effects of non-product-related brand associations on status and conspicuous consumption. J Prod Brand Manag 11(2):67–88

77. Omrane A, Bag S (2021) Determinants of customer buying intention towards residential property in Kolkata (India): an exploratory study using PLS- SEM approach. Int J Bus Innov Res 1(1):1. https://doi.org/10.1504/IJBIR.2020.10032274
78. O'cass A, McEwen H (2004) Exploring consumer status and conspicuous consumption. J Consum Behav 4(1):25–39
79. Parameswaran M, Whinston AB (2007) Social computing: an overview. Commun Assoc Inf Syst 19(1):37
80. Park H, Armstrong CMJ (2019) Will "no-ownership" work for apparel: implications for apparel retailers. J Retail Consum Serv 47:66–73
81. Pedersen ERG, Netter S (2015) Collaborative consumption: business model opportunities and barriers for fashion libraries. J Fash Mark Manag Int J 19(3):258–273
82. Perez ME, Castaño R, Quintanilla C (2010) Constructing identity through the consumption of counterfeit luxury goods. Qual Mark Res 13(3):219–235. https://doi.org/10.1108/135227 51011053608
83. Perry A, Chung T (2016) Understand attitude-behaviour gaps and benefit-behaviour connections in Eco-Apparel. J Fash Mark Manag Int J 20(1):105–119
84. Phau I, Prendergast G (2000) Consuming luxury brands: the relevance of the Rarity Principle. J Brand Manag 8(2):122–138
85. Piscicelli L, Cooper T, Fisher T (2015) The role of values in collaborative consumption: insights from a product-service system for lending and borrowing in the UK. J Clean Prod 97:21–29
86. PricewaterhouseCoopers (2015) The sharing economy. Consumer intelligence series. Available at https://www.pwc.com/us/en/technology/publications/assets/pwc-consumer-intellige nce-series-the-sharing-economy.pdf. Accessed 10 Jan 2019
87. Ratner C, Hui L (2003) Theoretical and methodological problems in cross-cultural psychology. J Theory Soc Behav 33(1):67–94
88. Rent the Runway (2015) How renting works. Available at https://www.renttherunway.com/how_renting_works#step-3. Accessed 10 Jan 2019
89. Reuters (2018). In sigh of times, luxury watchmaker Audemars embraces second hand. https://in.reuters.com/article/us-swiss-watches/in-sign-of-times-luxury-watchmaker-aud emars-embraces-second-hand-idINKBN1F80LN
90. Rivera FM (2018) Lifestyle for rent-luxury brands courtesy of the sharing economy. Available at https://www.europeanceo.com/lifestyle/lifestyle-for-rent-luxury-brands-courtesy-of-the-sharing-economy/. Accessed 10 Jan 2019
91. Rodrigues P, Brandao A, Rodrigues CS (2018) The importance of self in brand love in luxury consumer brand relationships. J Cust Behav 18:189–210
92. Roos D, Hahn R (2017) Does shared consumption affect consumers' values, attitudes and norms? A panel study. J Bus Res 77:113–123
93. Roos D, Hahn R (2017) Understanding collaborative consumption: an extension of the theory of planned behaviour with value-based personal norms. J Bus Ethics. https://doi.org/10.1007/s10551-017-3675-3
94. Roos D, Hahn R (2019) Understanding collaborative consumption: an extension of the theory of planned behavior with value-based personal norms. J Bus Ethics 158(3). https://doi.org/10.1007/s10551-017-3675-3
95. Roux D (2010) Identity and self-territory in second hand clothing transfers. Adv Consum Res 37:65–68
96. Roux D, Korchia M (2006) Am I what I wear? An exploratory study of symbolic meanings associated with secondhand clothing. Adv Consum Res 33:29–35
97. Roux D, Guiot D (2008) Measuring second-hand shopping motives, antecedents and consequences. Rech Appl Mark 23(4). https://doi.org/10.1177/205157070802300404
98. Ruvio A, Shoham A, Brencic MM (2008) Consumers' need for uniqueness: short-form scale development and cross cultural validation. Int Mark Rev 25(1):33–53
99. Ryan TA (2017) The role of beliefs in purchase decisions: a look at green purchase behaviour and altruism. J Res Consum 3:6–11

100. Sarigollu E (2009) A cross-country exploration of environmental attitudes. Environ Dev 41(3):365–386
101. Schwartz SH (1994) Are there universal aspects in the structure and contents of human values? J Soc Issues 50:19–45
102. Schwartz SH, Boehnke K (2004) Evaluating the structure of human values with confirmatory factor analysis. J Res Pers 38:230–255
103. Silverstein MJ, Fiske N (2005) Trading up: why consumers want new luxury goods - and how companies create them. Penguin Group Inc., New York
104. Stelling W (2018) The impact of the sharing economy on the luxury goods industry. Available at http://concinno.de/sharingeconomy/. Accessed 10 Jan 2019
105. Stern PC (2000) Toward a coherent theory of environmentally significant behaviour. J Soc Issues 56(3):407–424
106. Stern P, Dietz T, Abel T, Guagnano G, Kalof L (1999) A value-belief-norm theory of support for social movements: the case of environmentalism. Res Hum Ecolo 6(2):81–97
107. Stern PC, Dietz T, Guagnano GA (1998) A brief inventory of values. Educ Psychol Meas 58:984–1001
108. Strahle J, Erhard C (2017) Collaborative consumption 2.0: an alternative to fast fashion consumption. In: Strähle J (ed) Green fashion retail. Springer Series in Fashion Business, pp 135–155
109. Thøgersen J, Ölander F (2002) Human values and the emergence of a sustainable consumption pattern: a panel study. J Econ Psychol 23:605–630
110. Tilikidou I, Delistavrou A. (2004) The influence of the materialistic values on consumers' proenvironmental post-purchase behavior. In: Marketing Theory and Applications Proceedings of the 2004 American Marketing Association Winter Educations'Conference. Chicago, IL, pp 42–49
111. Triandis HC (1995) Individualism & collectivism. Westview Press, Boulder, CO
112. Triandis HC (1988) Cross-cultural contributions to theory in social psychology. In Bond M.H. (ed) The cross-cultural challenge to social psychology. Sage, Newbury Park, CA
113. Tsai S (2005) Impact of personal orientation on luxury-brand purchase value. Int J Mark Res 47(4):429–454
114. Tse DK, Wong JK, Tan CT (1988) Towards some standardized cross-cultural consumption values. Adv Consum Res 15:387–395
115. Tuncel N, Tektas O (2020) Intrinsic motivators of collaborative consumption: a study of accommodation rental services. Int J Consu Stud 44:1–13
116. Twitchell J (2002) Living it up: our love affair with luxury. Columbia University Press, New York
117. Veblen TB (1899) The theory of the leisure class. Houghton Mifflin, Boston, MA
118. Vigneron F, Johnson LW (2004) Measuring perceptions of brand luxury. J Brand Manag 11(6):484–506
119. Waight E (2013) Eco babies. Reducing a parent's ecological footprint with second-hand consumer goods. Int J Green Econ 7(2):197–211
120. Weil C (1999) Secondhand Chic: finding fabulous fashion at consignment, vintage, and thrift stores. Pocket Books, New York
121. Wiedmann KP, Hennigs N, Siebels A (2009) Value-based segmentation of luxury consumption behaviour. Psychol Mark 26(7):625–651
122. Wong NY, Ahuvia AC (1998) Personal taste and family face: luxury consumption in Confucian and Western societies. Psychol Mark 15(5):423–441
123. Wu DE, Boyd Thomas J, Moore M, Carroll K (2013) Voluntary simplicity. The Great American apparel diet. J Fash Mark Manag Int J 17(3):294–305
124. Xu Y, Chen Y, Burman R, Zhao H (2014) Second hand clothing consumption: a cross cultural comparison between American and Chinese young consumers 38:670–677
125. Yan RN, Bae SY, Xu H (2015) Second-hand clothing shopping among college students. The role of psychographic characteristics. Young Consum 16(1):85–98

126. Yeoman I (2011) The changing behaviour of luxury consumption. J Rev Pricing Manag 10:47–50. https://doi.org/10.1057/rpm.2010.43

127. Yeoman I, McMahon-Beattie U (2010) The changing meaning of luxury. In: Yeoman I, McMahon-Beattie U (eds) Revenue management: a practical pricing perspective, Chapter 6. Palgrave Macmillan, Basingstoke, UK, pp 62–85

128. Zaman M, Park H, Kim YK, Park S (2019) Consumer orientations of second-hand clothing shoppers. J Glob Fash Market. https://doi.org/10.1080/20932685.2019.1576060

129. Zhang M, Chebat J, Zourrig H (2012) Assessing the psychometric properties of Hofstede's versus Schwartz's cultural values of Chinese customers. J Int Consum Mark 24(5):304–319

Blockchain Technology in Apparel Supply Chains

Aswini Yadlapalli and Shams Rahman

Abstract The apparel industry is one of the fastest-growing industry employing millions of workers across the world. The majority of these people work in appalling working conditions resulting in the industry being called an industry with 'slave labour'. Moreover, the industry is known to be the second-largest polluter while consuming a significant amount of natural resources. Literature suggests that unsustainable practices are the result of a lack of visibility in the supply chain. Thus, promoting visibility in supply chains has become imperative in addressing the sustainability issues in the apparel industry. This book chapter presents how blockchain technology addresses the transparency issue and aid in the development of the sustainable apparel industry. The immutability and decentralisation properties of blockchain technology can promote sustainability. Benefits offered by blockchain technology implementation may vary among the supply chain members. For example, the use of blockchain creates ledgers of information that ensure the integrity and authenticity of organic cotton. Meanwhile, at manufacturers, the use of blockchain facilitates designers and merchandisers to share accurate information and collaborate remotely that assists to react faster to market trends. Moreover, it is used to monitor factory safety in its global supply chains. Despite the benefits offered by blockchain technology when implemented in supply chains, there are several challenges related to the implementation. In this chapter technological, organisational, and environmental (TOE) framework is used to examine the blockchain implementation challenges. Issues with compatibility, lack of skilled expertise, lack of clear government regulations, and privacy issues are some of the major challenges of implementing blockchain technology.

A. Yadlapalli · S. Rahman (✉)
Department of Supply Chain and Logistics, College of Business and Law,
RMIT University, Melbourne, VIC, Australia
e-mail: shams.rahman@rmit.edu.au

© The Author(s), under exclusive license to Springer Nature Singapore Pte Ltd. 2022
S. S. Muthu (ed.), *Sustainable Approaches in Textiles and Fashion*,
Sustainable Textiles: Production, Processing, Manufacturing & Chemistry,
https://doi.org/10.1007/978-981-19-0874-3_4

1 Introduction

The apparel industry with a turnover of US$3 trillion a year employs 40 to 60 million workers [15]. The majority of the apparel workers generally work in unacceptable conditions with fire safety hazards. Minimum wages which are approximately a fifth of the living wages, endless working hours, and forced labour led this industry to be labelled as an industry with 'slave labour'. On the one hand, the apparel industry is the largest polluter, next only to the oil industry. Moreover, the industry consumes a significant amount of natural resources including water and land. Hence the carbon footprint of the industry is high.

Studies suggest that the lack of visibility is one of the critical factors that contributed to unsustainability in the apparel industry. Transparency concerning how raw materials are sourced, apparel is produced, and the working conditions in which they were produced is crucial not only to boost efficiency [13] and eliminate counterfeit [29], but also enables consumers to choose apparel that is produced sustainably. Studies indicate that technologies play an important role in improving transparency in apparel supply chains and blockchain is one such technology. Recently, it has been marketed as a technology that provides transparency and traceability in supply chains through properties of immutability, decentralisation, and cryptography [4, 8].

In apparel supply chains, the use of blockchain replaces the centralised conventional distribution system with decentralised immutable ledger that connects the supply chain members with real-time information. Thus, the blockchain not only acts as a traceability tool but also promotes sustainability. The first apparel tracked using blockchain technology was made by Danish designer Martine Jarlgaard in 2018. Since then, other major fashion players have begun to explore the benefits offered by blockchain [7]. The objective of this chapter is to provide insights into how blockchain technology solves key issues relating to transparency and sustainability in apparel supply chains. This chapter focuses on the following areas:

- An overview of apparel supply chains is discussed in Sect. 2.
- Sustainability challenges exposed in the apparel industry are detailed in Sect. 3.
- Blockchain technology and its characteristics are presented in Sect. 4.
- Sustainability benefits offered by the implementation of blockchain technology in apparel supply chains are detailed in Sect. 5.
- Challenges of blockchain implementation in supply chains are discussed in Sect. 6.

2 Global Apparel Supply Chain

Since 2001, apparel exports have increased steadily from US$434 billion in 2001 to US$1,038 billion in 2019 which accounts for 5–6% of merchandise exports [13]. Every year apparel worth US$150 billion produced is consumed in locations far from where they are manufactured. In addition to the globalisation in manufacturing, apparel supply chains are dispersed globally from product design in one country through to raw materials sourcing from a different country, and, finally, sales across the world.

A simplified version of global apparel supply chains is shown in Fig. 1. Several supply chain members include but are not limited to are presented in each block. For example, upstream raw material suppliers in the figure exclude steel, minerals, and plastic suppliers. These raw materials are key in manufacturing accessories and components such as zippers and buttons that are critical supplies to apparel manufacturers in the supply chain. Meanwhile, these accessory suppliers are also not listed in the figure. At apparel manufacturer facilities operations may vary from an outsourcing model where the garment is stitched according to the buyer's specification from imported raw materials to the vertically integrated operations with a retailer designing, manufacturing, and selling the product. Irrespective of the operations, most of the apparel brands have consolidated their manufacturing to fewer countries [14]. Countries such as China, Bangladesh, and Vietnam have benefited most from this move and emerged as the leaders in apparel exporters with a share of 30.8, 6.8, 6.2%, respectively, in 2019 [24]. At upstream, export network organisations act as an intermediary between the retailers and apparel manufacturers by conducting procurement functions. At the downstream, retailer network consisting of flagship stores, department stores, etc., conducts the sales operations. In 2020, the top ten apparel brands gained an 11.4% market share [13].

In apparel supply chains, the two most noticeable trends worth examining are apparel production in low-cost countries and fast fashion. The retailer trend of outsourcing apparel production to manufacturers from developing nations has resulted in transparency issues with retailers having no information about the suppliers of raw materials. The level of visibility that a retailer had over the suppliers is illustrated by a pyramid iceberg model (see Fig. 2). It shows that tier—3, 4, and 5 manufacturers below the waterline are less visible in the apparel supply chain. Lack of visibility has provided an opportunity for apparel manufacturers to employ unacceptable practices.

The second trend in the apparel industry is a shift towards fast fashion where speed and control gained momentum. There is constant pressure on brands and manufacturers to reduce the time to launch the product. In some cases, the time to market was at about a month or even less. The increased number of collections has

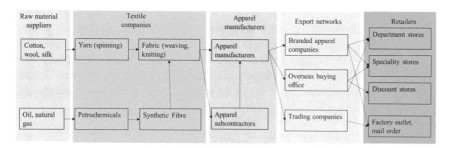

Fig. 1 Apparel supply chains. *Source* Authors

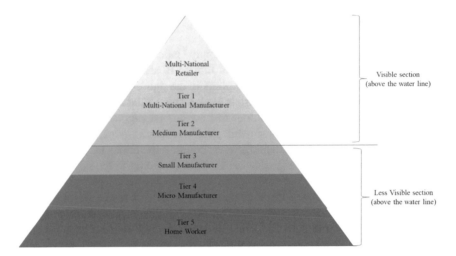

Fig. 2 Pyramid iceberg model in apparel supply chains. *Source* Gupta (2012)

minimised the price drop and product shortages which resulted in a reduction in losses and an increase in revenues for the fast fashion brands. It is expected that a 5% increase in revenue would lead to a profit increase between 22 and 28% and between 30 and 43% increase in market capitalization [13]. Despite the expectations of performance improvements, brands remain inefficient in terms of the end-to-end process. Instead of gaining efficiencies in the supply chain process by minimising the price reductions, overstock, and forecasting errors, brands are focusing on gaining efficiencies through pressuring manufacturers to decrease the apparel price. The pressures on manufacturers to shorten the time-to-market cycle coupled with the lower prices have resulted in inhabitable working conditions at apparel production facilities operating in countries with low regulations. The following section details the sustainability issues resulting from the current apparel trends.

3 Sustainability Issues in Apparel Supply Chains

Global spending on apparel every year equates to the GDP of the world's 126 poorest countries and it is expected that growth in apparel sales continues with an increase of 4–5% annually. An increase in sales coupled with the income growth of 4.3 billion Asians is expected to worsen the impact of the apparel industry on communities and the environment Sustainability issues in the apparel industry can be reviewed using the triple bottom line (3BL).

3.1 Economic Issues

In modern supply chains, the payment terms are benefiting the buyers while suppliers who sell the goods and services are taking the risks with the payments made to them after 30, 60, or 90 days after the sale. The apparel industry is one such industry with payment terms favouring buyers while forcing manufacturers to wait for the payments until the orders are shipped. In some cases, the situation with the manufacturer worsens as they are forced to accept payments anywhere between 30 and 150 days after the goods are shipped. This often leaves manufacturers with no working capital to increase their production capacity or to procure raw materials. This situation particularly threatens the existence of the small and medium enterprises as they do not have enough cash flows to pay wages, buy raw materials, and meet the operating costs incurred in producing the apparel that meets the future needs. The financial situation of the suppliers has worsened during the COVID-19 pandemic crisis. In terms of capital investment, COVID-19 caused some factories to delay or to decrease planned investment, including automation. Moreover, 40% of the brands/retailers have not made any commitments to make payments for completed orders leaving manufacturers in limbo [16].

To address the cash flow problems in the apparel industry, fashion factoring companies are developing a series of solutions. Businesses may obtain working capital through invoice factoring, accounts receivable financing, and purchase order financing. All these financing options require the management of documents and interactions with financial bureaucracy which requires additional resources. Moreover, apparel manufacturers are experiencing issues related to documentation resulting in delays at a rate of 66% of exports. This has resulted in money laundering risk about 53% of the time [9].

3.2 Environmental Issues

Fast fashion makes about 60% more in 2014 than it did in 2000 and only keeps them half as long [6]. The increased share of fast fashion comes at an environmental cost. The environmental impact of the apparel industry can be felt in the air, water, and soil. Globally apparel industry contributes to 10% of all carbon emissions which is more than the emissions from international airfreight and sea freight combined. If the emissions from the apparel industry continue, the share of the carbon emission could reach up to 26% by 2050 [20].

The trend of using clothes for less time means clothes end up in landfill. It is estimated that 85% of the textile produced goes into landfills contributing to 2% of the wastes in our landfills. The apparel industry produces over 92 million tonnes of solid waste every year which ends up in the landfill. On average one garbage truck full of clothes is dumped into the landfill every second [6]. This has become a serious concern as most of the brands started offering five collections in 2011 an increase from two collections in 2000. This number increases significantly for fast fashion brands. For example, Zara launches 24 collections per year, while H&M introduces between 12 and 16.

The apparel industry contributes to 20% of industrial water pollution and 35% of all microplastics in the ocean [20]. Chemicals discharged into water from washing the textiles reduce the dissolved oxygen in the water affecting the aquatic ecosystems. In addition, the apparel industry is the second-largest consumer of the world's water at about 1.5 trillion litres of water each year [6]. Farming cotton a primary raw material for the apparel industry is water-intensive. For example, over the last 50 years, water from the Aral Sea is used for cotton farming resulting in drying up the sea [20]. It is expected that cotton farming to cause extreme water scarcity in Asia by 2030.

3.3 Social Issues

Apparel brands are known to source from manufacturers operating in countries where worker rights are limited or non-existent. Often the brands move to different manufacturers in search of cheap labour. Though cheap labour is associated with low wages, many of the global brands are assuring that the workers are paid minimum wages. However, the problem is the huge difference between minimum wages and living wages. Living wage indicates the minimum amount required by the worker to fulfil the basic needs of the family. For example, in Bangladesh, the living wage required by the family is euro 259.80 per month whereas the workers are paid a minimum wage of euro 49.56 per month which is 19% of the living wage [6]. This means that the brands are paying workers 5 times less than what they need to support their family to live with dignity.

The minimum wages paid to workers are so low that they are willing to overtime. Often, apparel workers work for 14 to 16 h a day, 7 days a week. The normal working week for an apparel worker is 96 h per week [6]. The situation worsens when the manufacturer is approaching the deadline. Meeting production deadlines became a common scenario as the brands are launching more collections per year. Moreover, workers in the apparel industry usually work in facilities without proper ventilation, inhaling toxic substances including fibre dust and unsafe buildings. Long working hours in inhabitable working conditions impose serious health risks and may also expose workers to accidents. Incidents such as accidental fires, injuries, and structural damage to buildings became frequent occurrences at production facilities.

Despite knowing the factories are not safe, most of the workers were forced to work in unsafe factories as there are no unions to defend the workers' rights. In some exporting countries, government laws and regulations restrict the creation and joining the unions. In some cases, union members are physically attacked which discourages the other employees to form unions. For example, in Bangladesh, only 10% of the 4,500 garment factories have a registered union. The threatening behaviour has forced over 168 million children to work in the apparel and textile industry. In some cases, government initiatives force people and children to work in the apparel industry. For example, in South India under the Sumangali scheme, 250,000 girls work in textile factories for a minimum wage for three or five years. At the end of their employment tenure, girls are paid a lump sum that can be used for their weddings [14]. The wages paid to the workers and the working conditions have made the agencies such as European Parliament consider the apparel industry as an industry with 'slave labour'.

The environmental impact of the apparel industry and the working conditions in which the apparel is manufactured have branded the industry as unsustainable. Given the significant impact of industry on the environment, agencies such as the UN are seeking ways to promote sustainable fashion and to make the industry less harmful. Likewise, 56% of 64 sourcing executives responsible for a total sourcing value of over $100 billion agreed that responsible and sustainable apparel sourcing is critical [1]. However, the adoption of sustainability initiatives in the apparel industry is so slow that is not enough to counteract the damage caused by the growth of the apparel industry.

Lack of transparency is one of the critical aspects that contributed to slowness in promoting sustainability in the apparel industry. The ability to trace the apparel on how they are produced including the working conditions in which they are produced is crucial in promoting sustainability. Technology is often seen as a solution that can address all the issues including sustainability. Blockchain is a disruptive technology that assists in improving efficiencies in the apparel industry and promoting sustainability by observing every aspect of the product transformation throughout the supply chain [7, 17, 4].

4 Blockchain Technology and Its Functionality

Blockchain has revolutionised the way transactions are recorded and stored. Using blockchain data is stored on a decentralised network called blocks that are connected in a chain form. The chain is an electronic distributed ledger that users connect to by a network of computers. Cryptography is used to process and verify the transactions that are to be recorded on the ledger providing a validation mechanism. As the information is stored in multiple locations, blockchain offers benefits related to the absence of a single entity controlling the system.

In addition, blockchain provides a platform for information sharing thus promoting accountability between individuals and institutions with different priorities. Private blockchain with invitation-only members protects the members' identity and stores transactions. As the information sharing is automated human or machine errors are minimised. Moreover, the immutability property of blockchain protects data from altering thus preventing fraud related to data manipulation. In some versions of blockchain, the smart contract is added as an extra architectural property. The smart contract is a computer programme that can autonomously verify and execute the terms of contracts. Thus, with the smart contract, the blockchain can allow the self-enforcement of contracts. All the contract terms can be authenticated by the blockchain, validating that each party has performed the tasks it is responsible for [8]. Moreover, automating the data recording into blockchain updates the information with members in real-time and removes the laborious error-prone reconciliation processes with internal records.

Blockchain technology has the potential to affect the supply chains of industries ranging from financial and education services to manufacturing and food. Based on the development of this technology and its applications, the evolution of blockchain technology can be divided into three phases [11]. In Phase 1, blockchain usage is limited to cryptocurrency in applications related to currency transfers, remittances, and digital payments. In Phase 2, a smart contract that facilitates the formulation and the verification of contract execution is developed. In Phase 3, validation of data recording into the blockchain is automated through the Internet of Things (IoT) ecosystem. Figure 3 details the evolution of blockchain technology.

Blockchain Phase 1
• Introduction of blockchain
• Cryptocurrency for applications related to cash, remittance, and digital payment systems

Blockchain Phase 2
• Decentralised applications Dapps are introduced
• Greater scalability of applications

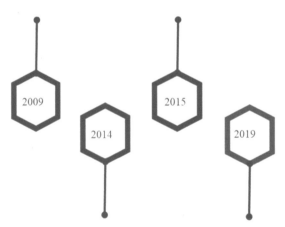

Blockchain Phase 2
• Privacy, smart contracts, and the blockchain tokens are added
• Applications-Ethereum and IBM-Maersk partnership

Blockchain Phase 3
• Automation of the validation process
• Inclusion of artificial intelligence into blockchain.
• Application CognitiveScale uses blockchain technology to store securely the results of AI applications built for regulatory compliance in financial markets

Fig. 3 Evolution of blockchain technology. *Source* Authors

5 Blockchain Benefits in Apparel Supply Chains

Technology is often considered as a solution for issues across all industries including the issues in the apparel industry. Digitalisation with the use of technologies such as blockchain in the apparel industry can change how supply chains are managed. This not only includes the introduction of new sales channels but also helps companies to adapt to new cost structures thus promoting the supply chain performance. For example, digital transformation introduces new sales channels such as click-and-collect and drive-through which attracts new customers. Technology is expected to become a new trend in promoting sustainable apparel supply chains. As a result, executives of some of the brands have made technology a core element of their company's strategy.

Particularly, the use of blockchain in the apparel industry provides supply chain members with real-time data that is immutable. Figure 4 shows how blockchain technology transforms the apparel supply chains towards circularity. The first proof of concept of blockchain in the apparel industry is used to track each step of the manufacturing process of a jumper through the Provenance app [7]. Since then, other major apparel brands have begun to explore the benefits offered by blockchain technology implementation in their supply chains. The following section presents the benefits offered by blockchain implementation to several members of global apparel supply chains.

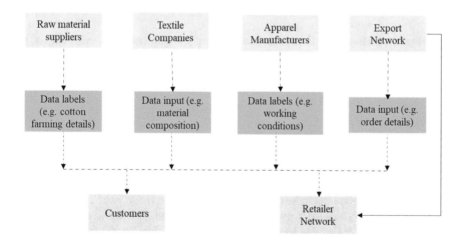

Fig. 4 Potential applications of blockchain technology in apparel supply chains. *Source* Authors

5.1 Raw Material Suppliers

More and more customers are interested in the details of raw material suppliers such as where, how, and in what conditions the raw materials are produced. Synthetic fibres, cotton, wool, and silk are the primary raw materials used in the apparel industry with a very distinctive production environment. Among all the raw materials, the use of cotton in apparel production with a share of 21% coupled with the adverse impact of cotton production on the environment has moved cotton production to the spotlight [3]. On average, cotton passes through six transactions before it is used in apparel production. To ease the complex paper trail between multiple parties that facilitate the movement of cotton from the farm towards the final consumer, blockchain can be used where the information such as prices, colours, weight, water consumption, fibre quality details, and more are recorded into the same ledger. The use of blockchain technology not only cuts the costs by replacing the individual ledgers of information at each stage of the supply chain but also promotes the increase in farm-gate prices. Thus, the transparency in the cotton supply chain through blockchain ensures the authenticity of cotton. Bext360, an agricultural blockchain is one of the blockchains start-up initiatives used to trace the cotton journey from farm to the gin where it is processed for textile use and from gin to consumer [17].

5.2 Textile Companies

The textile industry is considered as one of the most fragmented industries with over 10 members involved before the fabric is manufactured. The use of blockchain at textile manufacturers provides transparency on how the fibre was spun into a yarn which is woven or knitted into the fabric. The transparency in the textile industry through blockchain is used to provide assurance on the sustainability practices at the textile manufacturer and to avoid or protect the product from counterfeit. IBM Blockchain Transparent Supply (BTS) platform is a tamper-proof ledger that tracks the fabric from source to sales. Italian luxury textile manufacturer Piacenza has used the IBM blockchain platform to protect from counterfeit which is crucial for their customers [12]. Meanwhile, the TextileGenesis blockchain platform is known to promote a greener fashion industry. Global apparel brand H&M used TextileGenesis to trace the use of recycled polyester and certified responsible wool. Austria-based textile manufacturer Lenzing has extended the use of TextileGenesis to its customers in India, Bangladesh, and Pakistan to assure product quality [10]. This platform allows them to track the entire spectrum of fabric from the raw material to the sales. The technology facilitates brands to reach their sustainability targets and to gain a reputation as a manufacturer of sustainable textiles.

5.3 Apparel Manufacturers

Digitalization of spreading, cutting, bundling, sewing, pressing, and packaging operations in apparel manufacturing could increase efficiency and decrease lead times. This involves the development of a model to digitalise the operations and communicate the operations with supply chain members. Successful adoption of advanced tools and processes such as 3D product designing, virtual sampling, digital material libraries, and AI-supported planning requires a trusted information sharing platform such as blockchain. The use of blockchain facilitates designers and merchandisers to share accurate information and collaborate remotely thus assisting firms to reduce the time to market and sample costs.

In addition to the process efficiencies at apparel manufactures, blockchain is used to monitor factory safety in its global supply chains. For example, the introduction of ACCORD and Alliance safety agreements in Bangladesh meant all the manufacturers are undergoing inspections continually for safety compliance. In this context, blockchain can be used to record the inspection reports documenting instances of health and safety violations, child labour, or unauthorised subcontracting. Moreover, as the blockchain system is a distributed ledger it can act as a self-reporting infrastructure where workers record their experiences, thus giving them a real voice. As the auditor reports stored on the blockchain are timestamped that means no one can manipulate the results. Similar to social compliance, manufacturers are under pressure to adopt cleaner production practices. Partnership for Cleaner Textile (PACT) and Leadership in Energy and Environmental Design (LEED) are the cleaner production certifications given to apparel manufacturers. As the firms are increasingly adopting these practices, they can report them using the blockchain. There is a need for these firms to report and communicate the implementation of these environmental practices. In this context, blockchain can be used to document how environmental practices are implemented.

5.4 Export Network

Among all the supply chain members, firms in the export network need to communicate with many firms from other industries such as banks, carriers, trading organisations, etc. Exchanging secured information with the network members is crucial for an efficient export process. Blockchain enables information exchange via a secured decentralised network. Moreover, the optimum use of the technology is only possible when all the aspects of cross-border trade transactions are digitalized, from trade finance to customs, transportation, and logistics. Like the export network, transportation and logistics are seeking to implement blockchain technology as they are communicating with a large number of firms including banks and customs authorities [2]. Moreover, blockchain in financial institutions can digitalize and automate trade finance processes, in particular letters of credit, and ease supply chain finance.

As the information is exchanged instantaneously, blockchain reduces payment risks by providing fast, secure, and low-cost payment processing services. When implemented in procurement, blockchain replaces the conventional email order confirmation with an automated smart contract validation thus reducing human errors and misunderstandings in payment terms, returns, purchases, etc. Particularly, the technology becomes a powerful tool to facilitate small and medium entrepreneurs to participate in international trade by reducing trade costs and lowering the barriers to entry in international markets.

5.5 Retailers

Blockchain technology in the retail sector is opening up a world of exciting opportunities and efficiencies. In the retail sector, the technology enhances the interaction between retailers, customers, and suppliers. As the blockchain offers a ledger of immutable information on the flow of products throughout the supply chain, it assures that the apparel is produced ethically thus promoting consumer confidence. Hence retailers are increasingly seeking ways to deploy blockchain technology as a tool to exhibit their commitment towards sustainability. As retailers are selling products through multiple channels, they need to integrate all the sales channels into one omnichannel which is possible through blockchain where information is updated in real-time and shared with all the supply chain members. This means products bought online can be picked up in-store or collection centre. The seamless flow of information is only possible when the decision-makers integrate all the aspects of the supply chain such as production, buying, demand forecasts, channel allocation, and last-mile delivery. Availability of blockchain-as-a-service (BaaS) platform as an affordable, out-of-the-box solution drives the blockchain implementation by eliminating complex and expensive pilots. Through BaaS, a retailer can deploy blockchain technology across supply chain members who has the access to the internet browser [23]. Moreover, blockchain with other technologies such as AI and machine learning will process the data at a high speed to make crucial decisions.

5.6 Consumers

Lack of transparency in a company's supply chain with no visibility to suppliers is considered to provide a competitive advantage as no one is unable to build identical apparel. However, the trend has changed. Millennials would like to see where, how, and in what conditions the apparel is made as they do not trust the organisation's claims of sustainability. This is mainly due to the media attention that many brands have received for their environmentally unfriendly practices and slavery conditions. The tracking and tracing ability of blockchain technology has added more power to consumerism by enabling them to buy a sustainably produced product. Moreover, blockchain technology assists firms to transition towards a circular economy by encouraging customers to extend product use. For example, the MonoChain wallet allows consumers to maintain a record of the apparel details that they purchased. When the consumers realise the value of clothes, it encourages them to reuse, sell, or donate clothes thus increasing the life of the product [26]. This is very much in alignment with the backlash against fast fashion by the younger generation who are happy to reuse good quality products or purchase vintage pieces [21].

Implementation of blockchain technology among Asian apparel manufacturers would improve the sustainability targets of apparel production and the industry as a whole. Despite the huge attention, it is unclear how the technology changes the process of producing apparel hindering the implementation. Moreover, investments required for technology have resulted in implementation at a slow phase. Particularly, the Asian apparel industry has been slow in implementing the new technology and innovation due to low wages which outweigh the initial investments. This situation has worsened during the COVID-19 crisis.

6 Blockchain Technology Implementation Challenges

Despite the benefits offered by blockchain technology when implemented in apparel supply chains, there are several challenges related to the implementation. Successful implementation of blockchain technology to manage the products through the supply chain begins with the identification of challenges to be managed [22]. Some of the challenges such as financial costs, top management support, and governmental regulations are common to all the technology implementations in supply chains. To identify blockchain implementation challenges, a review of the literature on technology implementations in supply chains is conducted. Tornatzky and Fleischer's [27] technological, organisational, and environmental (TOE) framework is used to examine the blockchain implementation challenges. The technological category indicates the technological attributes, organisational category refers to the characteristics of the firm, and environmental category indicates the context in which a firm conducts its business. This framework is used as an underlying theory to study technological adoptions at organisational level. To extend the application of the TOE model at the supply chain level, inter-organisational relationship aspects should be integrated. Especially, the position of the firm in the supply chain, trust among the supply chain partners, and collaboration between the firms are ignored in the research. So, the TOE framework with the incorporation of inter-organisational factors is used to examine the challenges of blockchain implementation in supply chains. A framework on blockchain implementation challenges and its literature can be seen in Table 1.

Table 1 Blockchain implementation challenges

Challenge	Explanation	Sources
Technology-related challenges		
Relative advantage	Limited blockchain knowledge among the public hinders their ability to realise the potential benefits and relative advantages of blockchain technology in comparison with other technologies	Lielacher [19], Saberi et al. [22], and Sternberg et al. [25]
Compatibility	Current advancements in blockchain cannot handle all the functions of an organisation, thus blockchain needs to be compatible with the existing legacy systems	Kouhizadeh et al. [18] and Lielacher [19]
Complexity	The complexity associated with the technology is a concern for the implementation of blockchain technology. In addition, the complexity associated with the organisation's structure and supply chain is impeding blockchain technology implementation	Chang et al. [5] and Sternberg et al. [25]
Trailability	When organisations want to adopt blockchain technology in their businesses, they must know which blockchain fits their requirements. Blockchain testing mechanisms at trial phases would assist in examining the throughput of the system	Saberi et al. [22] and Wang et al. [28]
Observability	In any technology adoption, results from the existing applications assists in greater adoption of the technology. The lack of existing applications of blockchain technology in full scale impedes the projected greater adoption	Saberi et al. [22] and Sternberg et al. [25]
Organisational-related challenges		
Top management support	Blockchain is a relatively new technology with its first implementation less than a decade ago. As most of the applications are not very mature it is hard to gain top management support	Caldarelli, et al. [4] and Kouhizadeh et al. [18]
Technical expertise	Limited experienced team members on blockchain and even more limited expertise in assuring the technology implementation challenges the ability to implement the technology	Caldarell et al. [4], Chang et al. [5], and Saberi et al. [22]

(continued)

Table 1 (continued)

Challenge	Explanation	Sources
Financial Resources	Software required for the implementation of blockchain technology in the supply chain incurs high initial costs. In addition, the hardware required to integrate with the blockchain increases the setup costs. Thus, it is challenging for organisations to use blockchain technology	Lielacher [19] and Saberi et al. [22]
Firm Size	Larger firms are less agile compared to the smaller firms to implement technology. On the other hand, smaller firms are constrained by resources. The size of the firms is a factor that influences the successful implementation of blockchain technology	Caldarelli et al. [4] and Wang et al. [28]
External environment-related challenges		
Security	Blockchain network is vulnerable to an attack where miners can control 51% of the confirmation of the new transactions in the distributed ledger [19]	Lielacher [19] and Kouhizadeh et al. [18]:
Government Regulations	There are no regulations or standards on how the transactions should be recorded on the blockchain. In particular, information shared across the supply chain members using blockchain technology should be protected from data phishing	Kouhizadeh et al. [18] and Saberi et al. [22]
Industry Characteristics	Industry characteristics demonstrate the use of IT and its implementation. As block technology is in its infancy, it is not clear which industries are influenced to implement the technology	Saberi et al. [22] and Sternberg et al. [25]
Inter-organisational relationships related challenges		
Trust	The adoption of blockchain technology in the supply chain depends on the degree of trust between the supply chain members. However, building trust requires time and cooperative trading relationships	Chang et al. [5] and Wang et al. [28]
Privacy	Originally blockchains are designed to be accessible to all those who have made transactions over the network. However, to be adopted to supply chains and organisations blockchain needs to be customised. Customising blockchain to give rights for respondents to review only permissible parts requires planning and expertise	Lielacher [19] and Sternberg et al. [25]

(continued)

Table 1 (continued)

Challenge	Explanation	Sources
Information Sharing	Despite the importance given to information sharing, firms do not have confidence in information sharing with the supply chain members as this information is obtained by their competitors their businesses might be affected	Kouhizadeh et al. [18] and Saberi et al. [22]
Partner's power:	Powerful members in the supply chain may influence and force small firms into adopting IT solutions. Despite the partner's power, small firms are sometimes reluctant to adopt the technology due to the perceived high costs, or they are relying on traditional business operation routines	Wang et al. [28]

6.1   Technological-Related Challenges

Technological factors refer to the adopter's perceptions of technological attributes. Diffusion of innovation theory is used to provide a theoretical explanation of technological factors. Compatibility, complexity, relative advantage, trialability, and observability are the technology-related challenges impacting blockchain implementation.

Relative advantages: Relative advantage is the degree to which an innovation is perceived as better than the previous innovations. The adoption of blockchain technology in the supply chain depends on the relative advantages of decentralised system over the conventional centralised system. Despite the initial argument of increased traceability offered by blockchain among the supply chain members, it is not clear how this technology helps in enhancing the efficiency of data collection and transfer, streamlining the operational process, lowering the costs associated with product tracing, and improving the customer satisfaction.

Compatibility: Compatibility is the degree to which an innovation is perceived as being consistent with the existing values, past experiences, and needs of potential adopters. For the successful application of blockchain technology, a new management approach is required. As the blockchain cannot handle all the functions of an organisation on its own, it should be integrated with the current systems. Despite the need for integration with existing systems, current developments in the blockchain do not explain its compatibility and ease of integration.

Complexity: Complexity is the degree to which a technology is considered difficult to understand and use. Properties of blockchain technology offer an innovative way of information sharing among supply chain members which requires specific skills. Currently, resources are scarce with experience in blockchain technology implementation and usage. In addition, there is no protocol on blockchain implementation which further complicates the process.

Trialability: Trialability is the extent to which a technology is used on a limited basis. Innovations that can be trialable on a smaller sample provide confidence to the individuals who are considering it for adoption. When organisations want to implement blockchain technology in their businesses, they have to know which blockchain fits their requirements. In general, blockchain testing mechanisms at trial phases would assist in examining the throughput of the system. However, the initial use of blockchain technology has raised concerns on the issues of scalability. An increase in network members decreases the speed of transactions and increases the cost per transaction. In supply chains, scalability issues identified in the trialability phase limit the adoption as it requires planning on what aspects and products need to be traced in the supply chain using blockchain technology.

Observability: Observability is the extent to which the outcome of technological implementation is visible to others. If an individual can see the benefits of the technology implementation with ease, the more likely they are to adopt. In the context of blockchain implementation, it is hard to observe how blockchain technology is implemented in supply chains. The opportunity to observe the technology adoption is limited to firms that are a part of the process. In addition, the majority of the blockchain projects are controlled by management consultancy firms who restrict the information on the application of technology in supply chain traceability systems.

6.2   Organisational-Related Challenges

Organisation context refers to the characteristics and resources of an organisation that facilitates the technology implementation. In the context of blockchain technology implementation in apparel supply chains, top management support, investments, and technical expertise are organisational-related challenges.

Top management support: Top management acts as a change agent in sponsoring the project and ensuring that enough resources are available in implementing the project. The role of top management is crucial in all the stages from the start of the project through to the completion. Top management support is an important determinant in the successful adoption of innovative projects. Blockchain is a relatively new technology with its first implementation just over a decade ago. As most blockchain applications are not very mature, it is hard for the top management to realise the benefits of the technology. In addition, a lack of blockchain knowledge would inhibit management buy-in for blockchain adoption.

Investments: Software required for the implementation of blockchain need to be developed for a supply chain and is therefore expensive to purchase or develop. Moreover, the success of blockchain applications for tracing is influenced by the data. Hence automating the data collection became crucial which requires huge investment. In addition, there are costs involved in integrating the blockchain system with the existing legacy system. Thus, small and medium-sized organisations are far from realising the relative advantages of the use of blockchain in traceability systems.

Technical expertise: Blockchain technology is relatively a new field and is growing at a rapid phase that the professionals in the field are few in comparison to the demand. Due to an increased demand for professionals and limited supply, it has become a greater challenge to attract and retain blockchain experts. In particular, blockchain knowledge among supply chain members is very limited that results in competing for technical experts in a very limited supply. Huge costs involved in attracting and retaining the technical experts hinder the ability to implementation of blockchain technology among apparel supply chain members.

6.3   External Environment-Related Challenges

The external environmental category indicates the environment in which an organisation conducts its business. In the context of the apparel industry, security and government regulations are the major concerns that are influenced by the supply chain members.

Government regulations: Bitcoins based on blockchain technology have bypassed the regulation to avoid the inefficiencies in the conventional payment system. In times of crisis, conventional centralised systems act as shock observers resulting in less impact on participants. This could be one of the challenges of blockchain as it reduces oversight. Thus, the existing industry regulations should be adopted while implementing blockchain technology is a strong argument that blockchain applications should follow the existing industry regulations. In particular, there are no regulations on how the transactions should be recorded on the blockchain.

Security: Blockchains are susceptible to attacks of colluding selfish miners. Selfish mining is a critical issue because a cider or a group of coders controls more than 50% of the mining power. Blockchain network is vulnerable to a 51% attack where miners can control the confirmation of the new transactions.

6.4   Inter-organisational Relationships Challenges

The inter-organisational relationship category indicates the mechanisms used to manage the relationship between an organisation and its partners. In the study context, aspects such as power, trust, and privacy are the dimensions of inter-organisational relationships.

Partner's power: Partner power is an organisation's ability to influence another company to act in a prescribed manner. Power structure between business partners is highly correlated to blockchain technology implementation. Successful implementation of blockchain technology in supply chains is influenced by the capability of the dominating player. In supply chains, large retailers have the power to influence all their supply chain members towards blockchain implementation so that they can trace the origin of the products they are selling.

Trust: Trust is an important factor in explaining inter-organisational relationships. Supply chain members who trust each other will achieve the benefits of technology implementation. The adoption of blockchain technology in the supply chain depends on the degree of trust between the business partners. Trust is important for long-term cooperative trading relationships.

Privacy: Originally blockchains are designed as a public network where anyone can join and participate in the network. A disadvantage of a public blockchain is the openness with no privacy for data on the network. The use of public blockchain to share information among supply chain members raises concerns about the information leakage to the competitors. Meanwhile, the permissioned blockchain means access is given to a group of people with rights. Customising blockchain to give rights for respondents to review only permissible parts is a challenge as it requires planning and expertise.

Addressing these challenges is critical for blockchain implementation. The implementation of blockchain technology will provide opportunities to build sustainable apparel supply chains by providing transparency.

7 Conclusion

Globalisation and fast fashion trends along with emerging new rich in Asia have accelerated fashion consumption. Meanwhile, consumers became conscious of the adverse impact of the apparel industry on the society and environment thus exerting immense pressure on firms to implement social responsibility in their business practices. This chapter provides an overview of apparel supply chains and the sustainability challenges in the industry. The discussion highlights the lack of transparency as a major source for sustainability issues in the apparel industry which could be addressed by technological advancements. Blockchain technology with properties of immutability, decentralisation, and cryptography became not only a traceability tool but also a driver towards sustainability in apparel supply chains. This technology has added more power to consumerism by enabling consumers to buy a sustainably produced product. Despite the benefits offered by the technology, it is not easy to implement the technology. The challenges of blockchain technology implementation need to be examined to manage the implementation. In this chapter, a systematic review of the challenges is conducted by using the TOE framework. To address the challenges all the supply chain members including industry bodies should work with IT companies through forming consortia.

References

1. Berg A et al. (2019) Fashion's new must have: sustainable sourcing at scale McKinsey Apparel CPO Survey 2019. https://www.mckinsey.com/~/media/mckinsey/industries/retail/our%20i nsights/fashions%20new%20must%20have%20sustainable%20sourcing%20at%20scale/fas hions-new-must-have-sustainable-sourcing-at-scale-vf.pdf
2. Businessfactors (2021) Fashion and apparel financing. https://businessfactors.com/industries/ fashion-apparel-financing/
3. CO (2018) What are our clothes made from? https://www.commonobjective.co/article/what-are-our-clothes-made-from
4. Caldarelli G, Zardini A, Rossignoli C (2021) Blockchain adoption in the fashion sustainable supply chain: Pragmatically addressing barriers. J Organ Chang Manag 34(2):507–524
5. Chang Y, Iakovou E, Shi W (2020) Blockchain in global supply chains and cross border trade: a critical synthesis of the state-of-the-art, challenges and opportunities. Int J Prod Res 58(7):2082–2099
6. Charpail M (2017) What's wrong with the fashion industry? https://www.sustainyourstyle.org/ en/whats-wrong-with-the-fashion-industry
7. Chowdhary A (2019) Blockchain and its relevance in the apparel industry. https://apparelresou rces.com/technology-news/retail-tech/blockchain-and-its-relevance-in-the-apparel-industry/
8. Cole R, Stevenson M, Aitken J (2019) Blockchain technology: implications for operations and supply chain management. Supply Chain Manag Int J 24(4):469–483
9. Fashionating World (2018) Bangladesh rmg plagued by shipment delays and late document presentation. https://www.fashionatingworld.com/new1-2/bangladesh-rmg-plagued-by-shipment-delays-and-late-document-presentation
10. Gerretsen I (2021) How blockchain could make fashion greener. https://edition.cnn.com/2021/ 02/08/business/textilegenesis-blockchain-fashion-spc-intl/index.html
11. Gupta V (2017) A brief history of blockchain. https://hbr.org/2017/02/a-brief-history-of-blo ckchain
12. IBM (2020) IBM launches blockchain for high-end textile for transparency of the supply chain. https://www.ibm.com/blogs/research/2020/12/sheep-to-shop/
13. ILO (2021) The post-COVID-19 garment industry in Asia. https://www.ilo.org/wcmsp5/ groups/public/---asia/---ro-bangkok/documents/briefingnote/wcms_814510.pdf
14. ILO (2020) What next for Asian garment production after COVID-19? The perspectives of industry stakeholders. https://www.ilo.org/wcmsp5/groups/public/---asia/---ro-bangkok/---sro-bangkok/documents/publication/wcms_755630.pdf
15. Iglesias T, Haverhals E, De Wée T (2021) The fashion industry needs to break with its gender and women's rights problems. https://www.fashionrevolution.org/the-fashion-industry-needs-to-break-with-its-gender-and-womens-rights-problems/
16. Juststyle (2020) Survey finds 40% of fashion brands have not paid suppliers. https://www.just-style.com/news/survey-finds-40-of-fashion-brands-have-not-paid-suppliers
17. Knapp A (2019) This blockchain startup is partnering with fashion giants to make organic cotton traceable. https://www.forbes.com/sites/alexknapp/2019/03/04/this-blockchain-startup-is-par tnering-with-fashion-giants-to-make-organic-cotton-traceable/?sh=5fe48d571fd2
18. Kouhizadeh M, Saberi S, Sarkis J (2021) Blockchain technology and the sustainable supply chain: theoretically exploring adoption barriers. Int J Prod Econ 231:107831
19. Lielacher A (2018) Five challenges blockchain technology must overcome before mainstream adoption. https://www.nasdaq.com/articles/five-challenges-blockchain-technology-must-ove rcome-mainstream-adoption-2018-01-03

20. McFall-Johnsen M (2020) These facts show how unsustainable the fashion industry is. https://www.weforum.org/agenda/2020/01/fashion-industry-carbon-unsustainable-enviro nment-pollution/
21. Radocchia S (2018) Altering the apparel industry: how the blockchain is changing fashion. https://www.forbes.com/sites/samantharadocchia/2018/06/27/altering-the-apparel-industry-how-the-blockchain-is-changing-fashion/?sh=679bcad229fb
22. Saberi S, Kouhizadeh M, Sarkis J, Shen L (2019) Blockchain technology and its relationships to sustainable supply chain management. Int J Prod Res 57(7):2117–2135
23. Soni P (2020) What's next? 3 ways blockchain will shape retail in 2020. https://risnews.com/whats-next-3-ways-blockchain-will-shape-retail-2020
24. Statista (2020) Share in world exports of the leading clothing exporters in 2019, by country. https://www.statista.com/statistics/1094515/share-of-the-leading-global-textile-clothing-by-country/
25. Sternberg HS, Hofmann E, Roeck D (2021) The struggle is real: insights from a supply chain blockchain case. J Bus Logist 42(1):71–87
26. Textile Today (2020) MonoChain transforming circular fashion via blockchain trace-ability. https://www.textiletoday.com.bd/monochain-transforming-circular-fashion-via-blockc hain-traceability/
27. Tornatzky LG, Fleischer M (1990) The processes of technological innovation. Lexington Books, Lexington, MA
28. Wang Y, Han JH, Beynon-Davies P (2019) Understanding blockchain technology for future supply chains: A systematic literature review and research agenda. Supply Chain Manag Int J 24(1):62–84
29. fibre2fashion (2021) Blockchain is reshaping the world of textile & apparel. https://www.fib re2fashion.com/industry-article/8428/blockchain-is-reshaping-the-world-of-textile-apparel

Role of Designers in Ethical and Sustainable Design

Vandana Gupta and **Dhara Vinod Parmar**

Abstract Sustainability is a much used word in the last few decades leading to new innovations and developments in different sectors of the society. There are many alternative meanings of sustainability, but etymologically, the word 'sustain' means to keep something going or to provide support. In this perspective, we might think of sustainability as a way of preserving life and protecting the environment. In a more practical sense, sustainability can be described as any action or process that has little or no negative impact on the natural world or living organisms, including other humans. It's all about figuring out how to meet life's demands without harming society or jeopardising future generations. It would not be wrong to share how textile and other industries have included new raw materials in their basket along with the application of advanced technology which have shown positive results in environmental protection. A shift from fast fashion to circular fashion is another important initiative to meet the goal of sustainable development and protecting the mother earth. All stakeholders are playing their role to combat the situation of environmental pollution but it is important to understand the role of designers in the context of sustainability. The horizon of sustainability needs to expand by analysing the designs and their impact on society. Designing includes fundamental issues of practical decision making and is concerned with the values and standards by which human actions are categorised as good or bad. This chapter will discuss the ethical approach, designers of different fields should take to work towards the three pillars of sustainability. In addition, the views of different critics and authors will be presented leading to design thinking and moral responsibility at each stage of design development.

Keywords Ethics · Designers · Sustainability · Fashion · Design · Morality

V. Gupta (✉) · D. V. Parmar
Parul Institute of Design, Parul University, Vadodara, Gujarat, India

© The Author(s), under exclusive license to Springer Nature Singapore Pte Ltd. 2022
S. S. Muthu (ed.), *Sustainable Approaches in Textiles and Fashion*,
Sustainable Textiles: Production, Processing, Manufacturing & Chemistry,
https://doi.org/10.1007/978-981-19-0874-3_5

1 Introduction

The current scenario of environmental sustainability has brought the role of designers and importance of ethical designing at the forefront of different design industries. Design is all about improving life of the end user and it affects the way people interact with the world they live in; which is indirectly dependent on the type of decisions taken at early stage of design process by designers. The concern of environmental pollution is directly related to the type and quantity of products developed and created by the designers. They not only become part of the market but part of the daily life of human beings. Taking examples of some products such as synthetic fibres, plastic bags, etc., have although made life of human beings easy but they are also responsible for many adverse effects like health issues, air pollution, water pollution, etc. Apart from the introduction of such material in the market, the production of large amount of products has resulted in landfills, pre- and post-consumer waste. The increased purchasing power has made customers buy more than the required materials for their consumption and disposal of such products has harmed the environment globally. The technological advancement, increase use of digital tools and information technology have led human beings to dominate nature and be masterful in the field of good design [1]. Although different industries, organisations, brands are working towards the sustainable development but ethical side of the designing needs to be considered in the coming future to achieve the goal of sustainability.

Design forms an integral part of our culture. Traditional methods of creating products focused more on their daily purpose and multiple usage. Like one of the most famous textiles that is "Ajrakh printed" fabric of Kutch, Gujarat was made with all types of natural raw materials and takes more than 15 days to dye & print [1]. The fabric is used in constructing garments for different occasions and later with passage of time, due to reduced durability of the material, a new product is created for other purposes. Thus recycling the same material. Another such example is Kantha embroidery technique of Bengal. The old saris or dhotis are layered and embroidered into one fine piece of fabric to be used for other purposes [2]. In recent years, to combat the problem of environmental pollution researchers are taking advantage of agrowaste material that can be used to create natural dye by using chicken pea husk [3, 4] apparels by using silk waste and pineapple leaf fibre [5] and exploration and development of biomaterials. Looking into the concept of sustainability it is important to understand that each part and each step of product development contribute to the positive or negative effect of the product on environment and society. Taking an example, majority of the clothing is not designed to aid reuse or recycling at the end of its life but as suggested by Durham et. al., small modification in the design process of garment making, the damage of the fabric can be reduced when recycling of the clothing needs to be done [6]. This explains that design and design ethics plays a major role in putting the product in the classification of sustainability. Majority of literature shares the importance of focusing on sustainable approach for better life but very less work emphasises the role of designers in achieving sustainable development. This chapter talks about sustainable development, designers and their ethical contribution towards the three pillars of sustainable development.

2 Sustainable Development

The concept of sustainability dates back to ancient times and with the change in life style the concept and its meaning at all levels that is social, economic and environmental, have been changed. It has become a common language and trend in the twenty-first century and cannot be achieved in isolation therefore all the stake holders need to come together to meet the goal of sustainable development [7]. The consciousness at all levels of society, every industry, designer and consumer is the most important aspect [8]. Sustainability should focus on the management of resources to an extent that the future generation should be able to enjoy what the present generation is enjoying [9, 10]. Environment, society and economy are the three pillars of sustainability but looking at the core of these three pillars, all are connected with human wellbeing [11]. Sustainability is not limited to a thought but needs to be taken as a responsibility of admiring the past achievements, enjoying and appreciating the present and creating a place which provides sufficient basic necessities to each and every living organism present on this earth. Also, the decision making should be done by considering international, national, community, individual levels to reach the global sustainable development goals [12, 13]. Sustainable development is all about thinking for future while taking into consideration today's needs. It was first defined by Brundtland Commission in 1987 and the concept has undergone various modifications since then. According to the commission "Sustainable development is the development that meets the needs of the present without compromising the ability of future generations to meet their own needs". This explains that human activity should not be self-centred but a deep understanding of material, resources and people should be developed to provide better life to present and future generations [14]. It can also be considered as a vision of a better world where the health, security, peacefulness, economic opportunity all are integrated with biosphere protection. A definition adopted by the United Nations in its Agenda for development explains that a multidimensional approach is required to achieve higher quality of life and focus should be given on environmental protection, social and economic development which are interrelated components reinforcing sustainable development [15]. Today various organisations such as Sustainable Development Solutions Network—SDSN—of the United Nations launched in 2012, practitioners such as KPMG 2011 and academicians are working towards the implementation of the principles and objectives of sustainable development. The adaptation of the sustainable development approach can be seen in meeting the contemporary challenges of life and in protecting earth but the underlying principle remains the same [16].

3 Need for Sustainable Approach

It will not be wrong to say that humans have forgotten their relation with earth, instead of considering it as a place to live in; they are using it as a resource canter. In ancient times, people led their lives simply as members of natural ecosystems. Their activities depended on the availability of the bountiful resource of the Earth. Out of many planets, we the humans got an opportunity to live on earth which provided us with all the basic necessities such as trees provide us shade from the sun, give us oxygen to live, trees planted in forests act as the lungs of the world which helps human lungs to function at their best by giving oxygen and taking carbon dioxide from the environment. Apart from these basic requirements, earth is also a treasure house; this can be explained by taking an example of Phulwari ki Nal—Wild sanctuary of Rajasthan, which is a home of many medicinal plants and endangered species [17, 18].

It was the beginning of an age of mass production, mass consumption and mass disposal in which the environmental impact resulting from such human activities has increased significantly. Thus exploiting the earth for monetary/profitability and luxury comfort without any consideration of what we are giving back to the environment and where this comfort is leading. The kind of activities performed by humans, the suffering of the ecosystem can be examined in the food chain of animals and human beings. A simple example will be the use of fishnets leaving microplastic in water bodies which are consumed by animals like crabs and shellfish [19]. The wastewater discharged in the water beds and landfills have led to health hazards in humans as well as other species on this earth. All this has led to climate change resulting in numerous diseases, scarcity of basic amenities, disasters and psychological and physical disorders in human beings as well as other species [20].

With the intervention of technology in every sphere of life, the connection of human beings with nature has been reduced to a great extent. They are more restricted to man-made space and have less connection with the nature's assets such as sun, rain, changing weather, feel of natural flowers and plants. Their reduced empathy and attachment is the result of environmental pollution. In the coming decades, the natural beauty of earth will be only stored in the artificial machines and the future generation will be understanding the same only through digital format. It is also true that museums which play a major role in storing our past for future generations, will be displaying art and artefacts related to extinct flora and fauna. Approximately earth is calculated as 71% of water, 29% land and the maximum area is converted from green to barren land and coloured water. Rural and urban areas are converted into a garbage bin due to high amount of plastic, electronic and textiles waste. Within a few centuries we have seen a drastic change in the environment as the green belt is almost removed due to deforestation or burning and human-based construction as part of the earth. This change has also resulted in the migration of animals towards cities.

In the present scenario, in one part of the society the natural resources are exploited for recreational activities whereas the other part of society is starving for their basic requirements. Affordability by rich/upper class is greater than the lower class of society but it is important to note that the resources are limited. In the Covid-19 situation in India, there was scarcity of the most basic resource that is oxygen and the economic status could not save the rich as well [21]. Another point that needs to be taken into consideration is that there is a huge supply and demand of products which leads to malpractices. The products made are of low quality; using chemical-based raw materials such as polyurethane and polypropylene which reaches marine life and increases the plastic particle waste in the environment [22]. This urgent need of such a product was to protect human beings from the bad effects of Covid-19 but the consequences of the production and disposal of such small piece was not taken into consideration which has resulted in huge amount of pollution/health issues especially to animals. This is where conscious planning and holistic understanding is required by the producers, users, designers as well as innovators.

4 Designers and Their Role in Sustainable Development

When we talk about designers the list is too long such as there are fashion designers, interior designers, textile designers, costume designers, product designers, transportation designers, industrial designers, graphic designers, set designers and many more. Let us understand the meaning of the above-mentioned classification of designers, their role and skills required by designers of each design field.

Fashion Designer: Fashion by its very definition is about current custom or style; fashion designers express the zeitgeist, or spirit of the times in their work. Fashion is constantly changing and designers are expected to reinvent the design wheel every season. They deal with aesthetics and technical aspects of the Fashion and Apparel Industry. With increase in consumer awareness and sustainable materials, garments are developed and created in a more technical manner; providing additional comfort to the wearer or user like multifunctional garments, smart textile and apparels. The world is open for them as they can work in areas like film theatre and TV industry where costumes are required, apparel industry, with wedding planners as designers, home textiles and in sectors where indirect application of textiles are seen such as automotive, hospitality, medical, etc.

Interior designer: Interior design is all about research (materials, trends, buildings, and offices, human psychology, colour psychology and design theories), conceptual development, efficient planning (space planning, furniture planning and layout) and providing a relaxing environment for clients. They collaborate with architects, engineers and builders, filmmakers and visual planners to design the appearance and functionality of indoor (related) places. With changing lifestyles and rapid urbanisation, the scope of interior design has expanded dramatically, making it one of the most sought-after professions in India and worldwide today. With rising property prices, Indian homes are becoming smaller, and homeowners appear to have discovered a solution to the space dilemma in the form of interior design. People want theme-based interiors and styles to make their homes stand out, which could lead to a growth in interior design jobs in the future years. As an example: When we look around, we see a lot of high-rise residential and business structures. The designer or architect of a high-rise building is one, however that one ten-story building will most likely have 40 separate dwellings. As a result, we will technically require 40 interior designers.

Textile Designer: Whatever we are using has textile in one or the other way; it may be a bedsheet, a car seat belt, toothbrush, footwear, bandage and what not. The understanding of the textiles and its application in development of product especially through the traditional and advanced technologies used in apparel industry provides a wide scope to put your ideas into life.

Product Designer: Every product which fulfils the user's need and can successfully be mass produced is part of product design. Small concerns and issues in our daily life creates demand for new and innovative products starting from the toothbrush cover to a glass as well as the bed we are sleeping on. With more conscious consumers the industries like interior, glassware, automotive, medical etc., are looking for designers who can give solutions to the problems; keeping the aesthetics of the product in mind. The skills and understanding related to ergonomics, material, consumer needs, designing, industrial machinery will help them to see their ideas leading to the final product.

Transportation designer: With increasing population, environmental issues as well as customer demand for safe space have led to the need of innovation in the field of transportation design. With advanced technologies and materials, there is an important requirement of designers to provide the designs in which such material can fit to meet consumer demand.

Industrial Designer: Industrial Design is a course which offers knowledge about the concepts of designs and its execution of the same idea for any given product. The product can range from a mobile phone to computer, to technology to various different software and techniques associated with the respective product. The main aim of industrial design is to make products beneficial for the consumer as well as the manufacturer. The fresher in this field gets hired in various job roles and is paid simultaneously.

Graphic Designer/visual communicator: Visual communication is a method of graphically communicating ideas in a way that is both efficient and effective in conveying meaning. It's an important part of any content marketing plan. This is due to the fact that pictures can help elicit emotions in your audience, provide greater instances for your argument and much more. In fields such as graphic design, animation and media, both traditional and mainstream technology will be taught.

Set Designer: These industries globally are becoming more and more demanding and work areas are getting more challenging by the day. The young individuals who complete this degree course will step into the world with skills and confidence to choose their profession and work within various industries, and shall possess the skills to carve out a space of their own.

Although the design fields are different, each one creates a product by going through a design process which involves different stages such as define/brief, research, ideate, prototype, select, implement and learn. Each stage of product development requires questioning by oneself to deal with the constraints in the ongoing process. Questions such as who is the target audience? What kind of design solution is required? What will be the time and space of the design? Why design is required? How solution will be implemented? Who will be the target audience? All such brainstorming is important to reach the final decision making.

The first stage that is design is the stage in which the problem and target audience is selected/identified which allows precise understanding of the problem and constraints leading to development of appropriate solution. Research is all about investigating; it is about learning/finding something new or digging in the past through primary or secondary sources. Some designers research may include three steps; visual inspiration, gathering and sourcing information, understanding the market and consumers. Whereas some designers explain a different strategy to start the research process such as Malene Oddershede Bach (London Based fashion designer) explains that her research starts by exploring and understanding textiles, as it is the central part of all the collections [23]. The research stage reviews information such as the history of design problem, end-user research and opinion-led interviews, and identifies potential obstacles. Brainstorming sessions are involved which lead to ideas to meet the needs of the end users. Here the designer makes the decision to select the idea and prepare the prototype to be presented among the stakeholders before placing it in the market. Selection and implementation are important steps where the reviewed comments are discussed and the solution to develop the required product is made, thus implementing the final decision of development of product before delivery to client. Last but the most important one is the learning where the designer should look and collect the feedback from the customers or target audience and determine if the solutions/decision met the goals of the brief, leading to improvements in future. The design solution decided should meet maximum possible design brief as it is not possible to meet all the requirements of a brief within a single design due to different market segments [24]. Apart from satisfying the client/target audience, in this twenty-first century the three pillars of sustainability are part and parcel of the decision making process of each and every designers due to the market demand and current situation. Each field of design projects its own responsibility towards life. As design has huge influence on the entire life span of the product and its relation to the environment, designer should recognise how their design will impact the environment [25]. With increased population and spending power the market is exploded with products which are dangerous to environment and fall under unethical designs classification. These products are part of the market without any ethical consideration as they have more material purpose than any benefit to humanity as a whole [24]. A study conducted by Md. Mazedul Islain and Md. Mashur Rahman Khan suggested that with the help of Higg Index the product's environmental sustainability can be measured. The standard evaluation of the T-shirts in their study indicated that foreign branded products like S. Oliver, PUMA and Esprit showed higher score than the local branded products like Aarong and Yellow in terms of environmental sustainability [26]. Although the marketing strategy of ZARA (a Spanish fashion brand) of fast fashion has resulted in impulsive buying and increase in landfills. To be in the market other brands like H&M have also followed the same concept without analysing the repercussions [27]. The first copy of the designer products are also a huge concern, as the copied products are of low quality. Such products come in the category of counterfeit. This explains that some brands are trying to adapt themselves to the societal consciousness about environmental pollution and developing production processes that are not damaging to the environment but there are others who need to be educated and motivated to think about sustainable development [28]. Let us take a look at the sustainable approach taken by designers in the last few decades.

1. **Interior design:** It is a profession which involves designer's skills in creating and developing enclosed space leading to the settings within buildings that house human activity and their experience in these settings [29]. It includes moral responsibility of interior designers to protect, restore and preserve ecosystem by optimum utilisation of natural resources as this field of design involves their extensive use to create a home within a home. Keeping the aesthetic choices of clients intact and creating healthy, functional, comfortable and sustainable space by rationalising the resources and energy consumption can help interior designers to develop ethical designs. Avoiding harmful materials which might cause hazard to environment and encouraging and providing the sustainable solutions to their clients which are durable and require less maintenance and replacement should be the focus of any interior designer [30]. Ali Basim Alfuraty suggested that the process of promoting and enhancing the interior environments with sustainable materials is a shared responsibility of both interior designer and user [31]. In recent years, companies and brands such as Organoid, Granorte and Bolon are focussing on eco-friendly and sustainable materials best suited for interior design [31]. Surfaces developed by Organoid company by using organic materials such as flowers, plants, seeds and coffee can be used for wall panels or can be integrated into furniture designs. In addition to this, these materials have imbued scent and acoustic absorption properties which are best suited for personal spaces [32]. Swedish flooring designer BOLON develops rag rugs by using vinyl waste collected from the companies close to its production plant (25-mile radius) thus also reducing carbon emission [33]. Similarly Sahil Bagga and Sarthak Sengupta incorporate zero Kilometer Design concept by collaborating with local craftsman and use of local available materials in their interior projects [34]. Also, the amalgamation of ancient art and craft in developing interior spaces is another way through which these designers are creating inherently "green" or sustainable buildings. The traditional material and architecture was developed to meet the challenges of their respective environments such as thermal comfort, earthquakes, floods, etc. The material like terracotta, mud, wood, bamboo, etc., where used in line with nature and not against it [35].

2. Fashion Brands and Designers: Today's sustainable fashion design philosophy and movement promotes concepts like ethical production, working conditions, fair-trade, recycling, upcycling, sharing, renting, eco-friendly, green and circular fashion [35]. To meet the above shared philosophies the designer Stella MacCartney is using organic materials like organic cotton, regenerated

cashmere, solar panels for stores and recyclable materials for packaging. Her strategy is based on four fundamental pillars which respect nature, people, animal and include circular solutions in product development. Sandra Sandor the designer behind the brand Nanushka uses vegan leather to create bags instead of fur, animal skin etc., whereas Eileen Fisher believes in circular fashion and uses natural dyes and recycles old textiles to create luxury and sustainable clothing. Creating something from nothing is the practice of Katie Jones, a UK-based Knitwear eco-friendly designer which incorporates playful aesthetics with serious ethics; ensuring the developed designs addressing the issues related to landfill and over-consumerism [36].

3. Craft and Textile Designer: Shreya Jain known for her sustainable efforts has created a La Lairos, a luxe bedding brand which not only uses the eco-friendly and organic materials but also ensures zero waste in total supply chain [37]. Kvadrat, Denmark-based textile manufacturer, uses waste wool and cotton to produce high quality out-door upholstery (without fluorocarbon), window covering, rugs in their plant which is fuelled 100% by renewable energy [33]. Design intervention with traditional craftsman brings new life to the product and increases the possibility of sustainability. One such example is the products created by Ishrat Sahgal's design house, Mishcat Co. by using leftover materials of the silk sari industry. Each piece is well crafted by the weavers of Rajasthan and Uttar Pradesh who are responsible for creating exquisite carpets and dhurries by using different yarns and the artisans' style of weaving. Wills Vegan is a brand name of the designers "Will Green" whose work involves vegan and sustainable product design. Other brands such as Anokhi are working with craftsmen and share an honest relationship with them by providing them the livelihood in their own comfortable space. The brand works with organic cotton supplied by Pollachi near Coimbatore and natural dyes for development of their products. A more creative methodology towards sustainable designing is seen in the work of Doodlage where the designers and craftsman associated combine the waste pieces of fabric in an innovative silhouette; unknowingly adding character to their work [38]. Different brands and designers are working as per different philosophies like Christopher Raeban "remade"; Bottletop "along with using waste, they are providing livelihood to underprivileged"; Ka-Sha "not only saving environment by using old unused textile materials but also trying to motivate people to stop impulsive buying".

4. Product and Industrial Designer: Sector/industries such as hospitality, travel and tourism, interior, medical requires one or the other kind of product which are being used in very close contact with the human being and have daily usage. Therefore the product designers need to look and take into consideration the material, client, atmosphere, disposal of product, ergonomics, energy consumption, aesthetics and many more such variables while developing the design. Vestre is a sustainable furniture brand that focuses on zero emission through renewable energy source for production plant and transportation [39]. Whereas Mater, Danish design brand of furniture and light works with the principle of design, craftsmanship and ethics based on sustainability. Material sourced are

recycled aluminium collected from old car parts and bicycle wheels as well as felled Mango fruit tree after it stops bearing fruits. This makes the brand working at different pillars of sustainability [40]. There are brands such as Tala who is also promoting the environmental consciousness by donating part of their revenue for tree plantation programmes [41].

As discussed above, lots of work is being done in the name of sustainability or sustainable development but does the work meets the ethical requirement. Few questions are important to ask while considering ethical development such as the path followed in the development of product and services provided by the designer to consumer aligns with ethics or not? To understand the concept of ethical design which will help to meet the sustainable development approach, let us first understand the meaning of ethics.

5 Ethics

Ethics is a border term which talks about moral philosophy. It can also be understood as a discipline which shares what is good and what is bad. It provides us the theoretical guidelines to live life and perform the task and behave in a given situation. It deals with questions such as "is it correct to find happiness for oneself or one should also work towards others happiness". Its subject is the nature of ultimate value and the standards by which human activities may be considered right or wrong. It is a discipline that deals with both good and bad moral duty and obligation, moral principles, systems and theories connected to moral ideals that govern an individual or a group, professional ethics and guiding principles. While ethics and morality are often used interchangeably, there is a distinction in how they are employed. Morals normally imply a personal preference, but ethics usually implies universal justice and the question of whether or not a particular conduct is responsible [42]. Ethics is based on well-founded moral norms that dictate what humans should do, usually in terms of rights, obligations, societal advantages, justice or special qualities. A study conducted by sociologist Raymond Baumhart asked business people, "What does ethics mean to you?" People responded by saying

- "Ethics has to do with what my feelings tell me is right or wrong".
- "Ethics has to do with my religious beliefs".
- "Being ethical is doing what the law requires".
- ·"Ethics consists of the standards of behaviour our society accepts".
- "I don't know what the word means".

The response from people shows that many relate ethics with their feelings which cannot be part of ethics as feelings might deviate the person from doing what needs to be done in the given situation or what is ethical. If ethics is identified with religion then it is true that ethics would apply only to religious people which are not the case. High ethical standards can be set by religion which may become a source of motivation for ethical behaviour.

However, ethics is not the same as religion and cannot be confined to it. Following the law and being ethical are not the same thing. Most persons subscribe to ethical principles that are typically incorporated into the law. However, legislation, like feelings, can stray from ethical standards. Slavery laws before the Civil War, as well as the old apartheid laws in current-day South Africa, are egregiously clear examples of laws that vary from what is ethical. Finally, being ethical does not imply doing "whatever society permits". A society's ethics can be tainted at any time. Nazi Germany serves as an excellent example of a morally depraved civilization. While some people embrace abortion, the majority of people do not. If being ethical meant doing whatever society allows, one would have to come to terms with concerns that do not exist. So, what exactly is ethics? [43].

Ethics is a discipline of philosophy that analyses whether a human activity is right or wrong. This field of philosophy is particularly concerned with the questions of how a human being should act. What constitutes proper behaviour and a happy life? It is important to emphasise, however, that there is no universally accepted definition of ethics. This is due to the fact that ethics as a field is continually changing in response to changes in the social, cultural and political setting. The very least we can accomplish is characterise the nature and dynamics of ethics in relation to a particular time and place. For example, in Greek tradition, ethics is associated with the concept of the good life, therefore ethical investigation at this period was focused on figuring out what makes people happy. In reality, Aristotle Nicommachean presents not just a philosophy of happiness, but also methods for attaining it. Indeed, if we try to reconcile different points of view, we will have to define the relationship between doing what is right and being happy, which is a challenging challenge. We can't have an absolute definition of ethics for the same reason. Ethics have objective framework where it works as a good tool to find a way to difficult issues. It does not provide answers but helps to think in different line or direction. Or it can also be said that it helps to eliminate any kind of confusion and provide clarity about any issue which may lead to a conclusion decided by an individual, that is where the role of morality comes in the picture. As it is known as ethics and morality are similar but still have different meanings [43]. Many philosophers consider the two to be interchangeable. This is due to the fact that the former stood for the philosophy of proper behaviour and greater good. The latter refers to practice, or the correctness or incorrectness of human behaviour. To put it another way, ethics is the systematic investigation of the moral principles that underpin behaviour. Morality, on the other hand, is more subjective and prescriptive in character, involving sentiments, emotions and attitudes while contemplating a problem [44]. It teaches us what we should do and encourages us to do the correct thing. Morality is defined as an "end-governed rational enterprise" whose goal is to provide people with a set of norms that facilitate peaceful and communally satisfying coexistence by making it easier for them to live together and interact in ways that are beneficial to the realisation of the common good. As a result, ETHICS CAN BE DEFINED AS THE SCIENCE OF MORALS, WHILE MORALITY CAN BE DEFINED AS THE PRACTICE OF ETHICS.

1. Normative Ethics: This type of ethics is prescriptive in nature; it operates by establishing norms and standards that serve as a foundation for concerns such as what is right and wrong, good and terrible. As a result, it entails describing the behaviour, habits and responsibilities that one should learn and follow in order to directly share the behavioural repercussions that one will have over the other. Normative ethics is concerned with the development of standards or theories that outline how one should behave. As a result, normative ethics usually tries to come up with rules or theories that tell us how we should act. Ethics, which declare and define moral principles, can also be explained [45].

2. Meta ethics is a descriptive term. Summer claims that meta ethics seeks to explain the nature and dynamics of ethical principles, as well as the source and origin of moral truths and how humans learn and acquire moral ideas. This can be shown in these simple instances. As previously said, normative ethics establishes criteria for what is good and what is evil, or right and wrong, while meta ethics will inquire into "what do we mean by good or what do we mean by right?" When a moral philosopher addresses these issues, he or she is engaged in meta ethics. It's critical to recognise that meta ethics is concerned with the character and status of the claims we make [43].

3. Applied Ethics: Applied ethics emerged as the third major form of ethics during the evolution of the discipline. Applied ethics, as the name implies, is the application of ethical theories to determine whether ethical or moral actions are appropriate in a specific scenario. Individual moral issues or advisory capacity may motivate the practice of applied ethics. Using abortion as an example of an individual problem, where a person is torn between whether or not what he or she is doing is ethically proper, and an advisory moral issue, where a woman is suffering from an ectopic pregnancy and has no other option but to abort her foetus.

6   Designer and Design Ethics

6.1   Why Are Ethics in Design Important?

Consumers are paying greater attention to brands' moral values than ever before. Research shows 62% of consumers are attracted to brands that have strong, authentic ethical values. Consumers trust ethical firms because they believe the brand cares about their experience and they identify with the company. When brands adopt unethical practices, on the other hand, consumers lose trust in the brand, which can lead to lower brand loyalty and purchases. Every part of a brand's design, in the end, contributes to the message they are sending out. If a brand wants to develop content that reflects their company's beliefs, they should assess their design ethics on a regular basis.

Designers are the creative people who enjoy making things. Ethically they work for the betterment of society by identifying and solving problems. They are the one who brings new things into the world through their ideas, creativity and skills. Something that does not exist, they bring it to reality which is an extremely powerful act. There are different types of designers and their actions and thoughts are developed according to the industry they are working in. Some have to abide by the rules of the company, some can work as per their ethical and moral values and some need to act according to the situation without their will. As designers' job is not only to create new things, they have the responsibility towards the society by assuring what they are giving out in the world through their creations is made with ethical rules and principles. Designers should see design not just as an act of creation but as an act of choosing what to create and what not to create. Anything created without responsibility will definitely lead to destruction.

It is important to note, ethics is all about people, about the "other". It is concerned about other people rather than oneself. Therefore, thinking ethically means giving a thought about something beyond oneself [46]. As humans, we are considered as unique because we have been given the power/freedom to determine how to act. This thought is described by a Danish philosopher Soren Kierkegaard where he explains that falling from a cliff or not falling is our choice which determine what is right and what is wrong according to us. Ethics is also considered as an expression of a purpose and is a method of action. The designs are developed in constraints. This might not seem true to many but the designer should be able to understand that whatever they are creating must have some use and the product created should be according to that particular purpose which will involve number of variables such as size, time, strength, space, etc. [47]. Therefore it can be said that design is a process of deliberately creating a product to meet a set needs [23].

If design needs to be explained it is described as a process of developing a finished product or leading to a solution [48]. The purpose of design and how consumers perceive is well discussed in the book "The Design for Everyday thing" by discussing three types of tea pot design. The first design is more for a visual pleasure, the second design force the consumer to buy it because of its unique feature and the third design is made in such a way that the consumer can use it affectively and as per the purpose for which the design or product is developed. The first two products do not fulfil the performance criteria but still they are important for the consumer. The consumer can buy it keeping in mind its aesthetics or may be because of its performance characteristics. It is the responsibility of the designer to understand the good and bad effects of the product before putting it in the market platform. Designers should think of creating a product that client believes that he or she always wanted to have [49].

During the process of designing/creation two things play an important role that is design thinking and decision making at each stage of product development. Each decision taken up by the designer is an important one as it can affect the life of the customer using the product, environment in which it is going to be disposed off or society where it will be shared. It will not be wrong to say that design thinking is inherent, it is something which makes us humans different from other organisms. Different cultures, ages, gender, all sorts of people design one thing or the other such as a housewife changes the interior of her house time and again; a child creates a small piece of drawing or model [49]. Design thinking is required at every stage of life. But when it comes to designers' understanding of design thinking, it is well suited when a new idea is required to solve a problem [50]. On the other hand, decision making is a crucial process as it is a process which involves making choices [51, 52]. Designers think about different solutions and with the help of decision making process they reach a solution which they think is the best for the given/arisen problem. Let us consider an example of a product designed by a designer. A soap dispenser worked when used by white skinned person but not for black or dark skin person. This product showed its usability only for a particular group of people and because of poor design decision it has led to social nuances. Another example of technology and design decisions are 2010 Gadgetwise reported that the Xbox Kinect did not recognise the faces of dark-skinned gamers, Hewlett-Packard's uneven facial recognition software also had problems and Google Photos' auto-labelling system misidentified two black friends as "gorillas" [53]. The technology and design should be inclusive; it should have positive effect on society [54]. Also, as discussed earlier that design process involves many things such as technology, its use and selection is also an important part of design [55].

Design is nothing more or nothing less than the human beings' attempt to explain, manipulate and improve the environment. Design is not about how it looks, it is more about how it affects the individual or society. When the designer disregards the manipulation have on the environment, they are acting culpable. Any design process practised without giving it a thought about the consequences it may have during and after its use, without taking any responsibility for the developed design; the product falls in the category of destruction rather than creation.

Any design created is a team work, especially in today's era where multi-tasking, multi-skill is required to produce and market a product. For example, online marketing especially of textile and fashion products is a collaborative effort of fashion designer, craftsman, graphic designer, photographer, marketing expert, etc. Such team consists of smart people who are working on influencing the audience globally through visuals and make them impulsive to buy the product. If any decision on any level is taken which any of these designers think is not right and is a potential harm they should have the courage to share it. But the work culture might don't allow them to speak or the management do not give much concern to the same or the fear of losing job might stop the designer to say. Designers need to understand that they are the only one who are responsible for their work and should fear for the consequences of their work more than the consequences of speaking about it. A designer will be called a designer when the fear of causing harm is greater than the fear of getting into trouble or fear of taking responsibility.

Designers are working in two different formats; one who are making design process more predictable by using the tangible assets like information technology and digital gadgets. Such designs are the result of scientific procedure which stand for reason, logic and intellect but lose human psychic needed at the expense of clarity. On the other hand, feeling, sensation, revelation and intuition is all part of the design process which expresses romanticism, substituting sentimental passion for responses to human needs [56]. Victor Papanek, a designer and activist has shared a lot about good and bad design. He believed that the designers have ethical responsibility towards what they make. A book on "Design for the real world" shares number of concepts related to designers' responsibility. Having skills to design is not the only criteria to be a good designer but taking responsibility to to design for social benefit is also important as shared by Victor Papanek. Also, designers need to understand that responsibility is not the luxury but it is the core of what they do.

If the fundamental responsibilities of a designers are listed below given responsibilities can be considered as important.

1. Responsibility to the world they live in: Designers have the power to change things and have the freedom to do the right thing. People look up to them and follow what they say. Every single time a designer create or design a thing or an information/tool to make life simple, a lot is contributed by designer in terms of happiness, togetherness as well as towards humanity.

2. One designer represents all others: Designers have the responsibility to the craft they are working for. Their one decision, one move, one product affects all other designers' ability to perform their job and to contribute. Each and every designer is the representative of the other. Every time designer take on a job that is detrimental to the craft or behaving the way that undermine design, they are putting rest of us in the hole that we have to climb off. Also, the present generation of designers are learning from the work and experimentation done by the earlier designers therefore they too have the responsibility to document their work and share it so that the coming generation of designers can also be benefitted in terms of knowledge and experience and do not perform the same mistakes.

3. Responsibility towards clients: The main difference between a designer and artist is that a designer has clients and he/she works keeping in mind the problem/requirement of society whereas the artist does not have clients and they work for their own sake but do provide the aesthetic pleasure and mental satisfaction to society. First and foremost it is essential that the designer should select and choose the correct clients. One should look for people who have problem and actually looking for the solution of the same and in this process it is the utmost responsibility of the designers to provide the correct solution and do what is ethical.

4. Responsibility towards yourself: As a designer one should learn to say no. Keeping your view point and doing what is right is the important step towards ones own self-esteem. Designers are also known as gatekeepers, this means that they are responsible to share what needs to be shared. The reason why a person is called a designer lies in the choices they make. No one is in charge of their action but only designers themselves are and the portfolio they prepare defines the choices they have made in the course of time. Important to understand that it's us humans who need to act responsibly to make our earth a better place to live for the present and future. It needs to be believed that we (huge population of world) can make a difference with their small actions and positive consciousness.

Someone somewhere needs to stop this ignorance shown by many people and provide them the correct direction to actually achieve the goal of sustainability.

7 Conclusion

Sustainability is more than a trend; it is a comprehensive approach that begins with design and continues through all stages of the sourcing and manufacturing processes. Only when design optimises benefits to people and communities while limiting its own environmental impact can it be considered fully ethical. Sustainability must be at the heart of every brand, and it must guide their design decisions at all times. We hope that more products are developed that respect the environment and the rights of all those involved in their creation. Also it is important for designers to understand their responsibility and the cost of what happens when unethical design is built. When a design is developed, the designer is aware about its properties that is its harmful or useful aspects; seeing things going out of the door, which they themselves are dissatisfied with. Thus, breaking the trust of the users. It is important for designers to keep digging their conscious by asking what they want to focus and why give up when things get hard and what is the end of the created product. They need to understand with what we are surrounding ourselves and how they can encourage people to have a better life with the product designers design. In spite of the presence of unethical design products in the market there are designers who are putting efforts to match the ethical design production and consumption of products. They are working by considering design ethics in their process and products. An ethical design approach is based on questions rather than preconceived conceptions of right and wrong.

Also, because customers are the ones who drive any sector, changing their buying habits is the most effective way we can push ourself towards more sustainable future. This entails taking a deliberate approach that considers more than immediate enjoyment and considers long-term consequences.

References

1. Pathak S, Mukherjee S (2020) Entrepreneurial ecosystem and social entrepreneurship: case studies of community-based craft from Kutch, India. J Enterp Communities: People Places Glob Econ
2. Ghuznavi R (2004) The tradition of kantha and contemporary trends. Asian Embroidery, pp 131–142
3. Pandey R, Patel S, Pandit P, Nachimuthu S, Jose S (2018) Colouration of textiles using roasted peanut skin-an agro processing residue. J Clean Prod 172:1319–1326
4. Jose S, Pandit P, Pandey R (2019) Chickpea husk–a potential agro waste for coloration and functional finishing of textiles. Ind Crops Prod 142:111833
5. Hazarika P, Hazarika D, Kalita B, Gogoi N, Jose S, Basu G (2018) Development of apparels from silk waste and pineapple leaf fiber. J Nat Fibers 15(3):416–424
6. Durham E, Hewitt A, Bell R, Russell S (2015) Technical design for recycling of clothing. Woodhead Publishing, In Sustainable apparel, pp 187–198

7. (Gray 2010, p. 57). (Leslie Paul Thiele. 2016. Sustainability, Polity press, chapter 1, page 13). https://www.pmir.it/fileCaricati/1/Giovannoni%20and%20Fabietti%20(2013).pdf

8. (Leslie Paul Thiele. 2016. Sustainability, Polity press, chapter 1, page 13).

9. https://uwosh.edu/sirt/wp-content/uploads/sites/86/2017/08/Definitions-of-Sustainability.pdf. American Council on Education. One Dupont Circle NW, Washington, DC 20036–1193. Tel: 202–939–9452; e-mail: pubs@ace.nche.edu; Web site: http://www.acenet.edu; https://eric.ed.gov/?id=EJ796131.

10. Amutha K (2017) Sustainable practices in textile industry: standards and certificates. Sustainability in the textile industry. Springer, Singapore, pp 79–107

11. Mensah J (2019) Sustainable development: meaning, history, principles, pillars, and implications for human action: literature review. Cogent Soc Sci 5(1):1653531

12. https://uwosh.edu/sirt/wp-content/uploads/sites/86/2017/08/Definitions-of-Sustainability.pdf

13. (https://circularecology.com/sustainability-and-sustainable-development.html)

14. https://www.researchgate.net/publication/47697344_What_is_Sustainability

15. Tomislav K (2018) The concept of sustainable development: from its beginning to the contemporary issues. Zagreb Int Rev Econ Bus 21(1):67–94

16. Sharma S, (2019) Medicinal plant diversity in Aravallis. Int J Phytocosmet Nat Ingredients 6(1), 3–3. & Banu F, Sharma SK, (2016) A note on bryophytic flora of certain 'nals' of phulwari wild-life sanctuary. Indian J Environ Sci 20(1&2):53–57

17. Waring RH, Harris RM, Mitchell SC (2018) Plastic contamination of the food chain: a threat to human health? Maturitas 115:64–68

18. Weis JS (2020) Aquatic microplastic research—a critique and suggestions for the future. Water 12(5):1475

19. Almetwally AA, Bin-Jumah M, Allam AA (2020) Ambient air pollution and its influence on human health and welfare: an overview. Environ Sci Poll Res 27(20):24815–24830

20. (Oxygen shortage: many hospitals in Bengaluru stop admitting COVID-19 patients by Afshan Yasmeen Bengaluru, May 04, 2021, The Hindu https://www.thehindu.com/news/cities/bangalore/oxygen-shortage-many-hospitals-in-bengaluru-stop-admitting-covid-19-patients/article34475660.ece)

21. Fadare OO, Okoffo ED, 2020. Covid-19 face masks: a potential source of microplastic fibers in the environment. Sci Total Environ 737:140279

22. Seivewright S, Sorger R (2017) Res design for fashion. Bloomsbury Publishing

23. https://asimetrica.org/wp-content/uploads/2014/06/design-thinking.pdf

24. Islam M, Khan M, Rahman M (2014) Environmental sustainability evaluation of apparel product: a case study on knitted T-shirt. J Text

25. Atak A, Sik A (2019) Designer's ethical responsibility and ethical design. Univ J Mech Eng 7(5):255–263

26. Bianchi C, Birtwistle G (2010) Sell, give away, or donate: an exploratory study of fashion clothing disposal behaviour in two countries. Int Rev Retail, Distrib Consum Res 20(3):353–368

27. Morais CCDCP (2013) A Sustentabilidade no Design de Vestuário

28. Coleman C (2002) Interior design handbook of professional practice. McGraw Hill. http://aryacollegeludhiana.in/E_BOOK/home_science/Interior_Design_Handbook_of_Professional_Practice.pdf

29. https://www.researchgate.net/publication/321585173_Criteria_for_sustainable_interior_design_solutions

30. Alfuraty AB (2020) July. Sustainable environment in interior design: design by choosing sustainable materials. In IOP conference series: materials science and engineering, vol 881. IOP Publishing, p. 012035

31. https://www.adesigneratheart.com/en/interior-design-blog/sustainability-and-well-being/18-5-top-materials-for-a-sustainable-interior-in-2020

32. bolon.com

33. https://www.lifestyleasia.com/ind/culture/architecture/these-8-home-decor-brands-are-promoting-sustainability/

34. https://timesofindia.indiatimes.com/life-style/home-garden/traditional-indian-craft-inspires-modern-interiors-designers-for-sustainable-architecture/articleshow/78591173.cms
35. https://motif.org/news/top-sustainable-fashion-designers/
36. https://www.indiatoday.in/magazine/supplement/story/20150928-eco-warriors-ecological-creations-workshopq-the-retyrement-plan-jenny-pinto-733383-2015-09-18
37. kvadrat.com & https://www.dezeen.com/2019/06/04/sustainable-design-brands/
38. vestre.com & https://www.dezeen.com/2019/06/04/sustainable-design-brands/
39. mater.com
40. tala.com
41. https://www.britannica.com/topic/ethics-philosophy
42. https://www.scu.edu/ethics/ethics-resources/ethical-decision-making/what-is-ethics/
43. https://www.bbc.co.uk/ethics/introduction/intro_1.shtml
44. Kagan S (2018) Normative ethics. Routledge. Studies conducted by Immanuel Kant's shares lots of explanation on this concept. Johnson R, Cureton A (2004) Kant's moral philosophy
45. O'Neill J (2001) Meta-ethics. A companion to environmental philosophy, pp 163–176. Literature work related to Plato's provides a great insight what meta ethics actually mean and how it is related to every decision we make. Annas J (1981) An introduction to Plato's Republic.
46. (https://www.scielo.cl/pdf/arq/n49/art11.pdf)
47. https://appinventor.mit.edu/explore/sites/all/files/teachingappreceation/unit1/DesignUnit1.pdf
48. Cross N (2011) Design thinking: understanding how designers think and work, Berg
49. Luchs MG (2015) A brief introduction to design thinking. Design thinking: new product development essentials from the PDMA, 1–12.
50. Lunenburg, F. C. (2010, September)
51. THE DECISION MAKING PROCESS. In Nat Forum Educ Adm Supervision J 27(4)
52. Adman S (2017) 'Racist' soap dispensers don't work for black people. https://metro.co.uk/2017/07/13/racist-soap-dispensers-dont-work-for-black-people-6775909/
53. Vanderborght B, Okamura A (2020) United against racism and a call for action [Ethical, legal, and societal issues]. IEEE Rob Autom Mag 27(3):10–11
54. Hankerson D, Marshall AR, Booker J, El Mimouni H, Walker I, Rode JA (2016, May) Does technology have race? In Proceedings of the 2016 CHI Conference extended abstracts on human factors in computing systems, pp 473–486
55. Trends Desk (2017) Racist soap dispenser creates huge buzz on social media. https://indianexpress.com/article/trending/viral-videos-trending/this-video-of-a-racist-soap-dispenser-has-everyone-on-twitter-talking-4806625/
56 Papanek V (1988) The future isn't what it used to be. Design Issues 5(1):4–17

Model of Ethical Consumerism: A Segment Study on Handloom Reusable Menstrual Pads Adoption

Mukthy Sumangala and Pavol Sahadevan

Abstract Purpose: The purpose of the study is to identify the influence of proposed traits on female consumer's ethical consumerism toward reusable handloom menstrual pads and to provide managerial implications in the future. The paper provides a deep comprehension of how elements like product awareness, product attributes (product features and hygiene), consumer innovativeness, price sensitivity, past sustainable behavior, perceived obsolescence are related to females' adoption intention of handloom reusable menstrual pads. This study also examines different opportunities and challenges related to handloom reusable menstrual pads and their marketing challenges in the Indian Consumer context.

Method: Quantitative research was conducted among respondents aged below 20–40 years and above. Data was gathered through an online survey ($n = 102$) administered by web link that was distributed by e-mail as well as by posting on message boards focused on female customers. Information was collected with the respondent's knowledge, expressed willingness, and informed content. The data were derived from responses to structured questions including Likert scale statements and categorical questions representing the variables relevant to the study. The data analysis included univariate and bivariate statistical analyses—Percentage, ANOVA, and Regression.

Result: Through Regression Analysis, the paper identifies the positive and negative impact of product awareness, product attributes, consumer innovativeness, price sensitivity, past sustainable behavior, perceived obsolescence on females' adoption toward reusable menstrual pads. Additionally, female ethical consumerism was found to be significantly different between consumers of the different demographic segments.

M. Sumangala (✉) · P. Sahadevan
Department of Fashion Management Studies, National Institute of Fashion Technology (NIFT), Dharmasala, Kerala, India
e-mail: mukthy.s@nift.ac.in

P. Sahadevan
e-mail: pavol.sahadevan@nift.ac.in

Conclusion: This study contributes to the existing review of literature by discovering the effects of ethical consumerism on adoption of reusable natural-made eco-friendly menstrual pads. Managerial suggestions are provided to promote ethical consumerism among female consumers.

Keywords Ethical consumerism · Sustainable product adoption · Post-consumer textile waste · Handloom reusable menstrual pads · Mindful consumption

1 Context

Ethical consumerism can be seen as the changes to purchasing and consumption choices as a moral response to wrongs done in the production and distribution of goods and services. It helps in minimizing social and environmental damage and brings a positive impact to sustainability in a long run. In June 2010, the Government of India proposed a new scheme toward menstrual hygiene by a provision of subsidized sanitary napkins to rural adolescent girls. But there were various issues like awareness, availability, and quality of product, regular supply, disposal of napkins, reproductive health education and family support which needed simultaneous attention for the promotion of menstrual hygiene products. This study looks at the issue of product adoption not only from the marketing point of view, but also considers ethical and social values attached to it. Sustainable product adoption is now a widely recognized trend. Female hygiene product companies and social organizations around the world research on sustainable raw materials, processes and introduce marketing campaigns to promote mindful consumption among consumers and communities. They promote dialogue and are transparent about their social and environmental policies, practices, and impacts. There is a lack to address the needs of low-income rural women in terms of affordability of menstrual hygiene products in India. Handloom reusable menstrual pads are a good option due to its long product cycle, quality, and hygiene factors. This study is thereby, a step toward ethical consumerism through a better understanding of consumer behavior to adopt new innovative products like the handloom reusable menstrual pads. It is a challenge to position such products in the minds of consumer and promote acceptability and adoption. Different factors influence the consumer's adoption of reusable handloom menstrual pads like product awareness, product attributes (product features and hygiene), consumer innovativeness, social values, past sustainable behavior, and perceived obsolescence.

2 Ethical Consumerism and Related Concepts

2.1 How Do We Define Ethical Consumerism?

The concept of ethical consumerism has its beginnings back in mid-1990s, and consumers started to have a say in the manufacturing, sourcing and distribution of products. Both are forms of symbolic consumption because consumers consider not only individual but also social values, ideals, and ideologies [14]. Ethical consumerism alternatively called as ethical consumption, ethical purchasing, sourcing or shopping is a type of consumer activism. Lim [6] explained sustainable consumption issue from responsible consumption, anti-consumption, and mindful consumption perspectives. It is practiced through 'positive buying' where ethical products are preferred and 'moral boycott' that is negative buying and brand based purchasing. A good many number of authors (Smith et al. 1990) consider the ethical consumer as an evolution of green consumer. The 'green' or 'environmentally conscious' consumers (Kinner et al. 1974) are individuals with a general attitude of planet protection and conservation as well as ethical consumer behavior [11].

2.2 Ethical Consumerism and Sustainable Product Adoption

There has been a tremendous increase of consumer awareness on the impact of ethical behavior in context of sustainability and welfare issues, and this has brought exponential research into ethical consumerism. Ethical consumerism is no longer considered a niche [3]. Consumers have become aware that they can bring in change by changing the products they buy, even if it costs a bit more than the usual option available. As per Kirchhoff, C. [5] consumers by choosing certain products over others can embrace or reject certain environmental and labor practices and make other value claims based on the ethical values they hold. The consumers who are pro-environmental do not always buy sustainable products. Although consumers are regularly exposed to sustainable product messages, this does not always have an impact on actual adoption behavior. In the past, the relevant researches exploring the adoption behavior from the perspective of consumers which have been widely applied is Fishbein and Ajzen [1] theory of reasoned action (TRA). They stressed that an individual's decision to engage in a particular behavior is based on the outcomes the individual expects will come as a result of performing the behavior. In consumer behavior, adoption requires a sound attention that how they can be more influenced to adopt new trends (Workman et al. 1993). The adoption of handloom reusable menstrual pads can be seen a mindful consumption behavior of consumers. For they choose quality over quantity, which otherwise used to end up in landfills. Reusable menstrual pads are in trend and it is the healthier option, but years of conditioning and aggressive marketing have made us believe otherwise. The adoption of such product greatly depends on the product awareness and attributes like features and hygiene and other traits proposed in the study.

2.3 Female Ethical Consumer: From 'What We Are' to 'What We Want to Be'

Maslow's hierarchy of needs is a theory of motivation which states five categories of needs like physiological needs, safety needs, love and belonging needs, esteem needs, and self-actualisation needs. It distinctly explains why female consumers behave the way they do, and can relate to how the proposed traits influence the female ethical consumer on every level in buying, from the basic needs to the final needs level, like the self-fulfillment needs

- It's been ages since women secured her freedom to move, live and reproduce; hence it is a basic need undoubtedly. But even today, women of the poor socio-economic classes are not able to afford hygienic menstrual products. They spend their hard earned money on basic food and shelter, and resort to rags. The price sensitivity is higher in this level and the awareness about reusable menstrual product is tremendously low. The most important obstacles to ethical consumption in this level are the difficulties in obtaining information, product availability, and high prices of products [15] (Fig. 1).
- Menstrual products are basically a physiological need; women demand this need as their right to health and hygiene through revolutions. The World Bank and other non-profit organizations through analyses, collaborations with practitioners, and partnerships with governments are working to ensure that women are not limited by something as natural as periods [8].
- Women also need a job, safety, and the security it ensures, which influence buying more comfortable choices of menstrual products. They are price-sensitive by assuring the best quality for the paid price. In a study [13] revealed that most

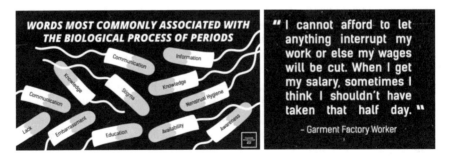

Fig. 1 Post shared by fashion revolution (@fash_rev) on Instagram, 2020

women in this level are not aware of the amount of plastic used in disposable menstrual products and suggests that higher product awareness will make them likely to choose products that are less harmful to the environment.

- Belongingness and love needs to ensure the consumption of healthier choices and can relate to the expression of who they are by the choices they make. Women always go back to the same ethical products if they get a sense of belongingness out of product they buy and reinforce the ethical buying behavior for a long period of time.
- Esteem is more important in the ethical consumption adoption. As the importance of sustainability and circularity reverberates, it's important for the modern women to be recognized as a thoughtful, mindful member of the society, their product adoption behaviors is shaped through this trend. Sustainable product adoption is such a widely recognized trend and women in the higher socio-economic levels easily adapts into it.
- Self-Actualization: 'I buy so I am' (Binet et al. 2018) explains the idea we are what we buy. When women buy reusable handloom menstrual pads, it reduces the post-consumer textile waste piling up which ends in landfills. Modern women consumer showcases her innovativeness by trying new menstrual products and choosing the best which send the message of ethicality and sustainability. It gives them a sense of acceptance in the society, confidence and satisfaction after purchase by not damaging the planet and its inhabitants.

2.4 Ethical Consumption Trends in Market and Handloom Reusable Menstrual Pads Adoption

In 2020, the Indian biodegradable sanitary napkin market will be worth US$ 550 million. In the years ahead, growth is expected to be at a CAGR of 33.4% during 2021 and 2026, which includes disposable and reusable menstrual pad segments. The growing knowledge of menstruation hygiene in India is fueling demand for biodegradable sanitary napkins. Furthermore, the use of high-quality and environmentally friendly raw materials to manufacture these pads is propelling the industry forward. Over the last several years, the Indian government has begun collaborating with a number of private companies and non-governmental groups to promote the use of biodegradable sanitary napkins, particularly among the poor and rural women. The Reproductive and Child Health Program, Eco Femme, and My Pad are just a few of these efforts. Aside from that, they're launching organic and chemical-free natural fiber options, which is helping to drive market growth (Table 1).

What is a reusable cotton pad?

Reusable cotton pads are pads that can be used and do not throw away during period. These pads are designed and worn in the same way as disposable pads. A reusable cotton pad is cleaned after each use and is ready to be worn again. Reusable cotton pads come in a variety of forms and absorbencies, just like disposable cotton

Table 1 Comparison chart of reusable menstrual hygiene products

Product attributes	Reusable products	
	Cloth Pads with/without Inserts	Hybrid Pads w/non cloth barrier
Price Range (Rs)	Rs. 85–400 per pad with an average of 250 Rs	
Reusable/Disposable	Reusable	Reusable
Compostable	Yes	Yes
Chemical/Plastic presence	Nil	Nil
Life Cycle Cost	Low	Low
Awareness	Low income clusters and high income	
Availability	Low income clusters and online	
Manufacturer	Ecofemme, Goonj, Gramalaya, Shomota, Soch, Uger, Saafkins	She Cup, Silky Cup, Moon Cup, V Cup, ALX etc

Source Census 2011 population data and International Institute of Population Sciences (2017). National Family Health Survey – 4, 2015 – 2016: India Fact Sheet

pads, however they are typically more absorbent. Most commonly produced disposable pads contains chemicals like styrene, chloromethane, chloroform, acetone, and toluene which can lead to skin irritations and major health concerns.

Reusable Menstrual products can be used multiple times whereas compostable disposable products are with high degree of compostable content.

- Life-span of 1–10 years resulting in minimal disposable impact
- Hygiene use requires care and maintenance; non compostable disposable can take 250 years to fully decompose.
- One time cost may be high but usually life time cost is lower than disposable
- The market has been segmented as bamboo-corn, cotton, banana fiber, and others. Bamboo corn currently dominates the market, holding the largest share

Key Areas marketers and manufacturers to be considered,

When choosing menstrual hygiene products, it is critical for market stakeholders (private sector and government) as well as customers to consider a variety of factors like awareness, access, usage, waste management, etc.

Due to the high competition and disposable pads monopoly unfortunately currently the reusable menstrual pads use is undermined by limited awareness and availability, which has not yet to been introduced to a majority of women. Marketers must be considering the factors like existing Myths and Taboos related to menstrual products, how consumers awareness related to product and brand can be improved, consumer educations on hygiene usage and maintenance of these product categories, initiating health seeking behaviors among female segment, etc.

A variety of menstrual product categories are currently available in Indian market, the majority of products reaching urban and rural consumers are likely to be disposable and non-disposable sanitary pads. It clearly indicates that markets has to analyze the accessibility parameters such as consumption costs, ease of accessibility, product choice availabilities like multiple products and brands, taxes and duties, involvements of decision makers, etc. in reusable product segment.

Marketers must have clarity on product usage and side effects, which should be communicated to the client segment. To meet consumer expectations marketers must focus on attributes like adsorption, fluid retention, non-allergic contents, etc. Also retailers have to consider the kind of raw materials and its ability to degrade, disposal frequency, amount of waste generated, ease of segregation, etc.

3   The Objectives of the Study

- To identify the influence of proposed traits on female consumer's adoption intention of reusable handloom menstrual pads.
- To study if product adoption differs between consumers with respect to their demographic profiles.

4   Hypotheses Development

Consumer innovativeness is defined as a tendency to buy new products rather than following familiar consumption patterns (Raskovic et al. 2016). Consumer innovativeness is often studied in the context of diffusion of innovation where consumers adopt new innovative product. Park et al. (2007) did a study on how three traits like fashion innovativeness, internet innovativeness, and materialism affect Korean consumers' attitude toward the purchase of foreign fashion items through e-tailors. According to Rogers (1962), consumers differ in terms of their distinct adoption manners when they are exposed to new services and products. Roger's classification of adopter categories are innovators, early adopters, early majority, late majority, and finally the laggards. Innovators, who adopt new products early, play an essential role in innovation diffusion (Cowart et al. 2008). Innovators act as a source of information regarding innovations (Morton et al. 2016). Innovators or early adopters of new products usually encourage and persuade others about the unique features of innovations. Hence, consumer innovativeness promotes new sustainable options like reusable handloom menstrual pads. Therefore, the following hypothesis is proposed.

H1:   Consumer innovativeness has positive effect on adoption of handloom reusable menstrual pads.

As per Advertising and Media insider report, adoption of disposable pads is less than 20% in India, whereas adoption of cosmetics like lipstick is significantly higher at 65%. The major barrier is affordability. Goyal (2016) in a study reported that around 70% of women in India say their family cannot afford the product every month. In such cases, it can be a sustainable option to buy reusable pads and develop the consumer base to rural areas. Voinea and Filip [16] in their paper found that price played a critical role in adoption decisions. Similarly, Braimah and Tweneboah-Koduah [2] demonstrated that price ranks ahead of sustainability concerns as a major influence in a decision making. To test, the following hypothesis is proposed.

H2: Price sensitivity has positive effect on adoption of handloom reusable menstrual pads.

Perceived obsolescence is when a product becomes obsolete in the minds of consumers mainly due to the effects of advertising, the new followers of products become obsolete to those who have not used them, almost forcing them to dump products that still work, or for the sake of model change, or to use products that others have. As per Neckel and Boeing [9] adoption influences the obsolescence process. This is due to the fact that it is from the adoption process that companies will have an idea of the product acceptance on the market, thus being possible to plan the launching of products with improved benefits. Consumers also recognize this factor, knowing that from the time of purchase of a commodity within a few months there will be a new product, with added benefits like hygiene and ease of use in the case of menstrual products. Perceived obsolescence reduces the value of older products in favor of a new product and influences the adoption. Therefore, the following research hypothesis is proposed.

H3: Perceived obsolescence has a positive effect on adoption of handloom reusable menstrual pads.

As per Verdugo [4] Sustainable behavior constitutes the set of actions aimed at protecting the socio-physical environment, and sustainable practices are regarded as positive behaviors, because they are universally recognized and valued, as ideals to meet and reproduce. Lindenberg and Steg [7] states that the search for well-being direct people toward the conservation of their socio-physical environment through getting involved in sustainable practices like adoption of ethical products. If the sustainable behavior is made a practice, they tend to do same for a long period of time. Consumers who have been participating in sustainable consumption practices are more likely to go with reusable cloth pads, hence the hypothesis can be stated as,

H4: Past sustainable behavior has positive effect on adoption of handloom reusable menstrual pads.

According to Ottoman [10], 'green products are typically durable, non-toxic, made from recycled materials, or minimally packaged. But there are no completely green products, for they all use up energy and resources and create byproducts and emissions during manufacturing stages, usage and even disposal'. Post-consumer textile waste is a big environmental problem even after major technology interventions. So green is relative, describing products with less impact on environment than their alternatives. This definition fits almost all recycled and reusable products. There is an explosion of the market for green products owing to a combination of consumer and industry infatuation with this type of product (Hopkins et al. 2009). The emergence of ethical trends also has an impact on consumption (Chen et al. 2008). Handloom menstrual pad is a long-lasting alternative to tampons and disposable pads. It is made of handloom absorbent fabric and can be re-used for up to 2 years if taken care of properly. Reusable pads are devoid of chemicals and that makes them an attractive option for women who are health conscious. It is not only healthier and safer to use, but for the customer, it is also more economical as reusable cloth pads overtime cost 1/8 the cost of disposables and better for the environment. Product attributes like biodegradability essentially lure pro-environmental consumers. High impactful attributes like the usage of biodegradable fibers like banana fibers in the pads is communicated effectively to persuade the consumer and hence the hypothesis is proposed as

H5: Product attributes have positive effect on adoption of handloom reusable menstrual pads.

Product awareness continues to drive positive change in market place. Awareness is a process of accepting and internalizing external information [12]. They suggested awareness to be understood as a major prerequisite for behavioral intention in cases of sustainable consumption. As per Park and Lee, growing product awareness outspreads the environmental issues to the awareness of alternatives during purchase, use and disposal stages of the sustainable product. Product awareness in this study means how much respondents perceive themselves to be aware of the environmental issues created by the product and the alternatives during stages of product lifecycle, also the strategies to deal with the issues. Their coping strategies range from eco-friendliness, mindful consumption, ethical consumption, creative consumption, re-use, slow-fashion, boycotting unethical brands, etc. Thus rapt an effect on the reusable product adoption and the following hypothesis is proposed

H6: Product awareness has a positive effect on adoption of handloom reusable menstrual pads.

5 Proposed Model

6 Research methodology

A quantitative study was performed with respondents aged below 20–40 years and above. Data was gathered via an online survey ($n = 102$), which was administered via a web link that was distributed via e-mail and by posting on female-oriented message boards. To guard against irrelevant or invalid results, a pre-test of 30 participants was conducted after the questionnaire was constructed. The required changes to the questionnaire were made in response to the respondents' questions. The respondent's expertise, expressed desire, and informed content were used to collect information (Fig. 2).

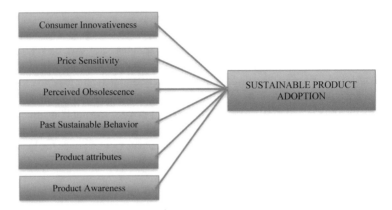

Fig. 2 Proposed conceptual framework

Measurement instrument and questionnaire design: The questionnaire is divided into two parts. Every Likert item will be graded on a 5-point scale, with 1 indicating strong disagreement and 5 indicating strong agreement. In the first session, a multi-item scale was created to quantify components such as price sensitivity, which was calculated using a Mujandir 5-item scale (2015). Park et al. suggested a 6-item scale, which was used to measure the variables consumer innovativeness and Product Awareness. Perceived obsolescence was measured using 3 item scale developed from Culture of consumption (2011). A 4-item scale was adopted to measure the variable Past sustainable behavior from Lang(2018). Product attribute was measured by adopting 3-item scale from Ottoman [10].

Questions about demographic variables were asked in the final session. Data were analysed using univariate and bivariate statistical tools including ANOVA, Correlation, and multiple regressions with SPSS (Fig. 3).

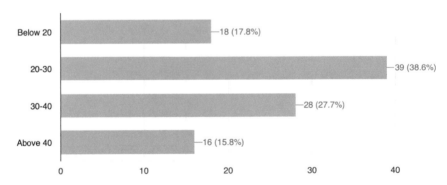

Fig. 3 Age

7 Data Analysis and Interpretations

Age:
The bulk of the respondents were between the ages of 20 and 30, with a fair representation of those between the ages of 30 and 40 (Fig. 4).

Occupational Status:
Majority of the respondents were students and a good presentation from full-time employees residing in Pan India (Fig. 5).

Annual Family Income:
Among the respondents, majority were having family income of above 10 lakhs, while almost an equal presentation from respondents having family income of 6–8 lakhs and 8–10 lakhs (Fig. 6).

Marital Status:
Majority (52.5%) were unmarried women while 44.6% of the respondents were married women.

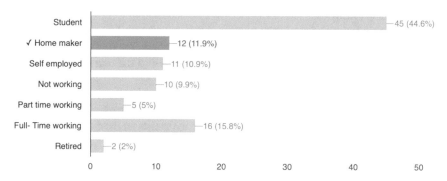

Fig. 4 Occupation

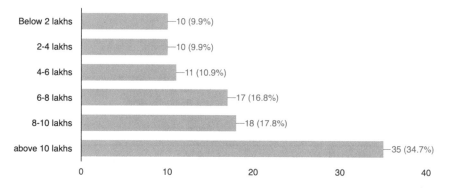

Fig. 5 Income

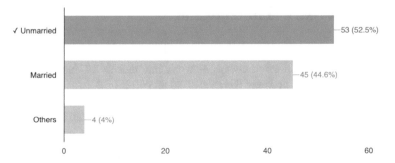

Fig. 6 Marital Status

Testing of Hypothesis

This section deals with the results of the statistical tests based on the hypothesis formulated in order to identify significant difference of demographic groups in respect of reusable pads adoption intention among female respondents.

Reliability Analysis:

Cronbach's alpha was used to measure internal consistency (reliability). It is most commonly used when we have multiple Likert statements that form a construct and we wish to determine if the items are reliable to measure that particular construct (Table 2).

Following are test results of reliability for particular constructs:

From the table, Cronbach's alpha value for every construct is more than 0.7, which indicates a high level of internal consistency for the construct used.

Multiple Regression Analysis:

Multiple regression Analysis was done (extension of simple linear regression), as there were more than two independent variables. Multiple regression is also useful to determine the over all fit (variance explained) of the model and the relative contribution of each of the independent variables toward the total variance explained. Here

Table 2 Reliability Test

Table reliability analysis			
Variables	Reliability statistics		
Consumer Innovativeness	Cronbach's Alpha	Cronbach's Alpha based on standardized items	N of Items
	.747	.749	6
Price Sensitivity	.866	.882	4
Perceived Obsolescence	.870	.874	3
Past Sustainable Behavior	.839	.843	4
Product Attributes	.853	.869	3
Product Awareness	.889	.893	3
Reusable Product Adoption	.824	.819	3

the dependent variable is Reusable Pad adoption intention and independent variables are consumer innovativeness, price sensitivity, perceived obsolescence, past sustainable behavior, product attributes, and product awareness. The results are represented through the following tables (Table 3).

A value of 0.758 for R, indicates a good value of prediction of reusable pad adoption intention by the independent variables. Furthermore, 54.8% of the variability is explained by the independent variables.

Table 4 shows that the independent variables significantly predict the dependent variable,

$F = 27.211, p < 0.05$. Thus, the model is a good fit of data.

$$Y = 0.940 - 0.042\, x_1 + 0.230\, x_2 + 0.083\, x_3 - 0.031$$
$$x_4 + 0.304\, x_5 + 0.288\, x_6$$

- A significant regression equation was established with respect to above variables with a p value less than assumed significance level (5%).

   From the Table, following are inferred:
- With $B = -0.042$, t value $= -0.619$ and $p = 0.010$, which is more than the assumed significance level 5%, we have to conclude that consumer innovativeness has negative and no effect on female reusable products adoption intention. Thus, we reject the H1.
- With $B = 0.230$, t value $= 3.414$ and $p = 0.001$, which is less than the assumed significance level 5%, we have the evidence to state that price sensitivity has a significant effect on reusable pad adoption intention of female consumers. Thus, we accept the H2.
- With $B = 0.083$, t value $= 0.903$ and $p = 0.036$, which is less than the assumed significance level 5%, we have evidence to state that perceived obsolescence has positive and highly significant effect on reusable pad adoption intention of female consumers. Thus, we accept H3.
- With $B = 0.031$, t value $= 0.303$ and $p = 0.016$, which is less than the assumed significance level 5%, we have to conclude that past sustainable behavior has positive and significant effect on reusable pad adoption intention of female consumers. Thus, we accept H4.
- With $B = 0.304$, t value $= 2.432$ and $p = 0.017$, which is less than the assumed significance level 5%, we have evidence to state that product attributes has a positive and highly significant effect on reusable pad adoption intention of female consumers. Thus, we accept H5.
- With $B = 0.288$, t value $= 2.195$ and $p = 0.031$, which is less than the assumed significance level 5%, we have evidence to state that product awareness has a positive and highly significant effect on reusable pad adoption intention of female consumers. Thus, we accept H6.

ANOVA Tests:

Table 3 Regression table 1

Table Model Summary for Multiple Regression Analysis

Model	R	R Square	Adjusted R Square	Std. Error of the Estimate	Change Statistics				
					R Square Change	F Change	df1	df2	Sig. F Change
1	.758[a]	0.575	0.548	1.80907	0.668	103.784	6	361	0

[a]Predictors: (Constant), Consumer Innovativeness, Price sensitivity, Perceived Obsolescence, Past Sustainable Behavior, Product attributes, and Product awareness

Table 4 Regression Anova table

Table ANOVA[b] table for Regression Model						
Model		Sum of Squares	Df	Mean Square	F	Sig.
1	Regression	416.501	6	69.417	27.211	.000[a]
	Residual	307.638	94	3.273		
	Total	724.139	100			

[a]Predictors: (Constant), Consumer Innovativeness, Price sensitivity, Perceived Obsolescence, Past Sustainable Behavior, Product attributes, and Product awareness
[b]Dependent Variable: Reusable Pad adoption intension (RI)

Table 5 Regression table

Table Coefficients[a] table for Multiple Regression						
Model		Unstandardized Coefficients		Standardized Coefficients	t	Sig.
		B	Std. Error	Beta		
1	(Constant)	.940	1.34		.701	.010
	Consumer Innovativeness	-.042	.067	−.047	−.619	.053
	Price sensitivity	.230	.067	.332	3.414	.001
	Perceived Obsolescence	.083	.092	.093	.903	.036
	Past Sustainable Behavior	.031	.104	.038	.303	.016
	Product Attributes	.304	.125	.295	2.432	.017
	Product Awareness	.288	.131	.255	2.195	.031

[a]Dependent Variable: Reusable Product Adoption Intension (RI)

Table 6 Testing of Anova Table

Table 4.4.1- DEMOGRAPHIC VARIABLES WITH RENTING INTENTION (ANOVA)								
SL NO.	Dependent Variable	Demographic Variables		Sum of Squares	df	Mean Square	F	Sig.
1.		Age	Between Groups	13.456	10	1.346	1.531	**.141**
			Within Groups	79.079	90	.879		
			Total	92.535	100			
2.		Annual Family Income	Between Groups	56.593	10	5.659	2.166	**.027**
			Within Groups	235.190	90	2.613		
			Total	291.782	100			
3.		Occupation	Between Groups	83.632	10	8.363	1.474	.161
			Within Groups	510.665	90	5.674		
			Total	594.297	100			
4.	Reusable pad Adoption Intention	Marital Status	Between Groups	4.798	10	.480	1.578	..126
			Within Groups	27.361	90	.304		
			Total	32.158	100			

To test the significant difference in reusable pad adoption intention among female consumers of different demographic profile (Annual Family Income)ANOVA test was employed and results of which are given below (Table 6).

Also Post hoc **Tukey HSD - Multiple Comparisons,** there is statistically significant difference in reusable pad adoption intention between female consumers having annual family income of 2–4 lacs and 6–8 lacs ($p = 0.003$), as well as between 6–8 lacs and Above 10 lacs ($p = 0.03$). There is also significant difference between consumers having family income of 2–4 lacs and 4–6 lacs ($p = 0.033$).

Differences were statistically significant (p value < 0.05) for the reusable pad adoption intention between customers of different annual family income. Thus, the hypothesis is accepted.

8 Findings and Suggestions

The purpose of this study was to examine the relationships between personality traits, including consumer innovativeness, price sensitivity, and perceived obsolescence along with the past sustainable behavior and female consumers intention to adopt reusable cloth pads. Study results illustrate the following findings:

- Among the respondents, majority of the females were youngsters, either students or full time employees of the 20–30 and 30–40 age groups. Most of them were unmarried and among the respondents, majority was having annual family income above 10lakhs.
- It was found that the reusable cloth pad adoption intention differs between annual family income. Reusable product adoption is a form of lifestyle that can be weaved into anyone's life with more disposable income. Thus, it differs among consumers having different annual family income. Hence it becomes important to educate all levels of consumers that it is a better option to invest in environmental friendly reusable cloth pads even at higher price. For they can choose quality over quantity, which otherwise end up in landfills.
- Consumer Innovativeness has negative and no effect on the reusable pad adoption intention of female consumers. It is thereby confirmed that consumers have distinct adoption manners when they are exposed to new products; hence have some restrictions while adopting them. Other factors like ease of use of the product and planned obsolescence can be relevant in innovativeness context which needs to be investigated in further studies. Reusability for cloth pads means washing again and again which minimizes the ease of use to the consumer and using it for a longer period delays the obsolescence, which in turn can have an effect on innovativeness.
- Price Sensitivity and Perceived Obsolescence were found to have positive effect on reusable pad adoption intention of female consumers. No previous research was found that has investigated the perceived obsolescence with regards to reusable cloth pad adoption intentions. Perceived obsolescence reduces the value of older products like disposable pad and tampons in favor of a new product like reusable cloth pads. This supports the marketers who can foresee and adapt to meet the needs of consumer.
- Results confirmed that past sustainable behavior contribute significantly to the prediction of consumer's reusable pad adoption intention. Consumers who have more positive beliefs about sustainable consumption or those who are more confident in their sustainable behaviors have stronger intention to adopt reusable pads. Furthermore, consumers who have had prior experience of participating in sustainable consumption practices are also more likely to go with reusable cloth pads.
- Also Product Attributes and Consumer's Product Awareness have positive effect on the reusable pad adoption intention. Current pandemic and lock-down situations urge marketers to communicate the added benefits of using reusable handloom menstrual pads and the environmental issues created by disposable pads. Since its reusable, there is no need to go out and stock disposable pads every month. This will boost the cloth pad usage, production, promote localization and eventually lead to sustainable societies. Product attributes like biodegradability essentially persuade the pro-environmental consumers. Marketers need to

communicate this information to prospective consumers in a timely manner to enhance the adoption process. They can also introduce marketing campaigns to promote mindful consumption practices.

With respect to theoretical applications, this study adds a unique contribution to the body of knowledge pertaining to reusable pad adoption intention of female consumers' in context of India by identifying the role of consumer innovativeness, perceived obsolescence, past sustainable behavior, and price sensitivity. This research is in its beginning stage and can provide basis for future researchers to justify looking at different factors more deeply in future studies.

9 Conclusion

Although reusable pad adoption had created a lot of interest, its acceptance is still in its infancy and popularity. The current study has confirmed a positive effect of perceived obsolescence and a negative influence of consumer innovativeness on the reusable pad adoption intention of female consumers. In addition, the results also highlight the positive effect of past sustainable behavior and product awareness on reusable pad adoption. This study has punctuated how personal factors like consumer innovation, price sensitivity, and perceived obsolescence along with past sustainable behavior can affect the reusable pad adoption intention. Profiling consumers based on personal factors would help marketers and retailers to customize marketing strategies to encourage greater adoption of reusable cloth pads. This study has few limitations. Firstly, women living in India were chosen which may limit the generalizability of the study. Future studies may examine across broader samples. Furthermore, longitudinal research which explains the model in association with behavior would complement the study findings.

References

1. Ajzen I, Fishbein M (1977) Attitude-behavior relations: a theoretical analysis and review of empirical research. Psychol Bull 84(5):888–918. https://doi.org/10.1037/0033-2909.84.5.888
2. Braimah M, Tweneboah-Koduah EH (2011) An exploratory study of the impact of green brand awareness on consumer purchase decision in Ghana. J Mark Dev Competitiveness 5:11–18.
3. Caruana R, Carrington MJ, Chatzidakis A (2016) "Beyond the attitude-behaviour gap: novel perspectives in consumer ethics": Introduction to the thematic symposium. J Bus Ethics 136(2):215–218.
4. Corral Verdugo V (2012) The positive psychology of sustainability. Environ Dev Sustain 14(5):651–666. https://doi.org/10.1007/s10668-012-9346-8
5. Kirchhoff C (2016, May 17) Ethical consumerism. Encyclopedia Britannica. https://www.britannica.com/topic/ethical-consumerism
6. Lim WM (2017) Inside the sustainable consumption theoretical toolbox: critical concepts for sustainability, consumption, and marketing. J Bus Res 78:69–80.

12. Lindenberg, steg (2007) Normative, gain and hedonic goal frames guiding environmental behavior. J soc issues 63(1):117–137.
8. Menstrual hygiene management enables women and girls to reach their full potential (2018, May 25). World Bank. https://www.worldbank.org/en/news/feature/2018/05/25/menstrual-hygiene-management
9. Neckel A, Boeing R (2017) Relation between consumer innovativeness behavior and purchasing adoption process: a study with electronics sold online. Int J Mark Stud 9(3):64. https://doi.org/10.5539/ijms.v9n3p64
10. Ottman J, Books NB (1998) Green marketing: opportunity for innovation. J Sustainable Prod Des 60(7):136–667.
11. Papaoikonomou E (2012) Sustainable lifestyles in an urban context: towards a holistic understanding of ethical consumer behaviours. Empirical evidence from Catalonia, Spain. Int J Consum Stud 37(2):181–188. https://doi.org/10.1111/j.1470-6431.2012.01111
12. Park S, Lee Y (2020) Scale development of sustainable consumption of clothing products. Sustainability 13(1):115. https://doi.org/10.3390/su13010115
13. Peberdy E, Jones A, Green D (2019) a study into public awareness of the environmental impact of menstrual products and product choice. Sustainability 11(2):473. https://doi.org/10.3390/su11020473
14. Uusitalo O, Oksanen R (2004) Ethical consumerism: a view from Finland. Int J Consum Stud 28(3):214–221. https://doi.org/10.1111/j.1470-6431.2003.00339
15. Uusitalo O, Oksanen R (2004) Ethical consumerism: a view from Finland. Int J consum stud 28(3):214–221.
16. Voinea, Filip (2011) Analyzing the main changes in new consumer buying behavior during economic crisis. Int J Econ Pract Theor 1(1):46.

Traditional Textiles Going Local and Global

Paul-Henry Dubois Athenor, Nina Hintzen, Natsuho Igarashi, Mukta Ramchandani, and Ivan Coste-Manière

Abstract As the innovation within the textile industry grows, there is an increase in demand for fabrics that are universally in trend and timeless. Often when local textile designs are reinterpreted globally, almost negligible credit is given to the local communities involved. Making it even more relevant today to maintain the authenticity and transparency for a global sustainable luxury and fashion industry. In this chapter, we highlight the challenges of the local traditional textiles going global, its impact on the fashion industry, and how these can be addressed to solve the pertaining issues for sustainable production and consumption.

Keywords Sustainable luxury · Sustainable fashion · Textile industry · Traditional textile · Fashion globalization · Fabric production · Fashion plagiarism · Sustainable design

1 Introduction

The textile industry is an integral part of our economy of high value and has undergone several developments in recent years. Regarding the textile market, one can see that the industry is represented internationally, but there are still significant areas

P.-H. Dubois Athenor
226, Chemin des Coudoulets, 84200 Carpentras, France

N. Hintzen
Gustav-Poensgen-Straße 49, 40215 Düsseldorf, Germany

N. Igarashi
4 Bis Avenue de Provence Imperial Bay, 06590 Théoule-sur-Mer, France

M. Ramchandani
Hombergstrasse 76, Olten 4600, Switzerland

I. Coste-Manière (✉)
SKEMA Business School, Université Côte d'Azur, 60 Rue Dostoiveski, 6902 Sophia Antipolis Cedex, France
e-mail: ivan.costemaniere@skema.edu

© The Author(s), under exclusive license to Springer Nature Singapore Pte Ltd. 2022   123
S. S. Muthu (ed.), *Sustainable Approaches in Textiles and Fashion*,
Sustainable Textiles: Production, Processing, Manufacturing & Chemistry,
https://doi.org/10.1007/978-981-19-0874-3_7

of concentration. When it comes to the largest textile producing countries, China is ranked first regarding production volume (52% of textile output in the global share in 2019), followed by India (6.9%) and the United States (5.3%). Besides these top textile producing countries, it can be stated that the textile industry is also fast increasing in other developing countries, due to greater investments [4]. On consideration of the share in world in 2020, it becomes obvious that again China was the premier global textile exporter by far, with a share of 43.5%. The European Union is listed as second with a world share of 18.1%, followed by India, with 4.2% [33]. This data underlines the high impact of globalization on the textile industry.

Even in early years, globalization has reduced trade barriers between countries, formed trading systems and has driven the countries' connectivity [9]. Globalization drives the economic strength of developing countries because fashion exports are a great driver of the creation of jobs. As a result, the employment rate of developing countries significantly increased. Besides these good aspects of globalization, the risks must also be considered. As a result of globalization, the issue of outsourcing in emerging economies increases due to the significantly lower costs. But working conditions are often hazardous to health and wages are too low, which represents a major risk for the population of developing countries. This is a call to action to support decent work. This could be for example carried out by international trade agreements [28].

While analyzing the textile industry, the consumers have an enormous impact. It can be outlined that the continuously changing consumer behavior significantly influence the textile industry. The globalization led to the fact that consumers actually strive for a greater selection of products which has resulted in a faster product cycle and shorter product life in the marketplace [5]. People place a lot of importance on an environmentally friendly lifestyle nowadays, so the manufacture of products must be oriented humanely and environmentally friendly [20]. Reflecting this on the current trends in the textile industry, a rising customer consumption regarding sustainable and natural fibers can be stated based on an increased environmental responsibility. This trend will further boost the textile industry in the future and contribute to its growth [15]. At this point, it is especially important to continuously address the consumer's mindset and behavior regarding over-consumption and fast fashion [27].

This chapter deals with the question: *Which dynamics make a traditional textile influence a local environment, and then become reinterpreted in a global market?* This question is structurally answered within this chapter, by providing and combining theory and analyzing practical business case examples. When considering a "local textile" and a "global textile", it is assumed that the roots of a traditional textile can always be traced back locally. While the focus of this report is clearly on making a traditional textile global, the local aspects must also be illuminated to remain competitive and maintain local values traced back to culture, tradition, and history.

1.1 Traditional Textiles Going Local

On closer examination of local textiles, this chapter seeks to answer the question: How can a textile, which may be traditional or be local? The local point of view always forms the base and starting point of a traditional textile. The local aspect of a textile is especially important as it embodies deep information concerning culture, tradition, and history. An understanding should be created that the local aspects of textiles are an upstream step of the analysis of the global viewpoint. Therefore, it should be referred to the foundation of a local textile, including cultural and physical aspects.

The history of textiles is tightly linked to the history of cultures. The common point of the Iranian carpet and the Japanese Kimono is the heavy cultural footprint they refer to. In both cases one is entering an entire universe, with a long local history and tradition. Textiles give us a lot of information about cultures, as they are a main produce of a cultural and technical environment. Moreover, it's not complicated to imagine that a traditional textile is obviously born as a local production link in the first ground, with some very basic conditions such as available materials or specific needs.

Textiles are deeply dependent on the available resources and knowledge, as well as being produced and consumed first locally. The textile is, by essence, representing a local production. During the middle age, more precisely during the twelfth century, Southern European countries were making tapestry and clothing out of cotton, linen, and wool. Whereas, in the Northern European countries, cotton was absent. Due to the climatic conditions of the Northern countries, cotton couldn't grow. This very difference between the two produces reveal the local anchoring of textile, with all the benefits it implies for a local market.

Another element underlining the original location of a fabric is the way it has been produced. In most cases, for very local and traditional textile, the making process is different everywhere. As per the article by Anna [6], Peruvian fabric makers were using a large range of different typical techniques, adapted to a local need or a particular ornamentation. In the same range of ideas, but concerning a totally different period and area, the studies of Dominique [7] on the *l'Arte della lana*, an anonymous treaty, probably written by a wool and sheet merchant at the very beginning of the fifteenth century. The author gives a very precise and technical information about processing, to produce an ornamentation textile. We can find the name of these textiles "Sarge" mentioned in Italian merchant's books added to a material or local indication such as "Linen Sarge" or "Reims Sarge", hence this proves the technique has been adapted to different material and locality in order to fit with a local need.

Not only textiles reflect some local cultures, but the local cultures could be built around specific fabric. In order to understand this fundamental aspect of a local fabric, we analysed it through the specific example of the Lyon's silk fabrics. This tradition has been implanted in Lyon in the sixteenth century. King Francis 1st decided to give a special monopolistic authorization to the city to import silk from Italy, in order to produce satin silk fabrics for royal palaces, diminishing the prices and boosting some local economy. In the idea, the fabric is entering twice a local content. Firstly, it permitted a growth of French interest in the silk commerce because weavers were importing some silk material from Italy and added value in the French town; secondly when fabric for the palaces was made, only few French merchants were allowed to buy it and resell it to the Royal court. In the end, from the processing to the consumption, this textile became a local business for the Lyon economy.

Moreover, the footprint of this fabric didn't stay only a good business for Lyon, but an all culture was built around that. According to a 1788 government's inquiry, almost 39,000 people were working in the silk business in Lyon, including weavers and merchant. Comparing this number to the city population of this time could give a good idea of the anchoring silk fabric in the city. According to different census, the population at this age was between 94,000 and 120,000. This means that silk business was employing a fourth to a third of the local population. This ratio complies with a number given by the *Scientific American* in 1868, announcing 120,000 workers for 300,000 population. Consequently, as we understand a word was specifically created in the French language to call Lyon's weavers. This term, "Canut" is still used to call silk weavers as well as other words, and it still reminds of the specific location of Lyon.

The silk workers also have a great implication in the political field. The "canut" term has a special resonance in the French history. This term reminds us of the 1831 and 1834 silk workers revolution, who's better known to be a social revolution. Without entering in the historical details, which is not part of this chapter, this revolution had consequences on all working rules of an entire country. According to Fernand Rude's writings, and on the many reports of his work, this revolution didn't lead to a chaotic situation or to massacrers, but to a quite calm and ordered city, at least for the 1831. This is explainable with a simple fact, the main point was not to revenge on some people but to get some working conditions and laws, that will benefit all workers. All the point of this revolution was, for these workers to get a kind of common regulation on the price of their work pieces.

Walking in the actual town, you could see that satin silk left another major trace in this city. In the northern part, built on the Croix-Rousse slope, the architecture is direct in heritage of the silk period. This neighborhood was built at the very beginning of the nineteenth century to respond to the explosion of the satin silk demand. Josette Barre's article (soiree lyonnaise et habitat. Typologie des immeubles de la Croix-Rousse vers 1830), help us to perceive the classification of architectures, relying on the production or commercial use of these buildings. The commercial buildings, for example, were built on the lowest part of the slope, easily accessible from the business neighborhoods whereas, the workshops buildings were built in the north, after the slope on the plateau. Writing this article, miss Barre knew that thirty years later after her study period, in 1856 the "Palais de la Bourse", French name given to the stock exchange building, would be built on the main avenue leading to the slope. The fabric footprint is entering here in a third dimension of the local life, the geographical organization of a city, relying on political choices and decisions.

1.2 Traditional Textiles Going Global

After showcasing the impact of the local environment of textiles, the following sections highlight the global environment. The main question here is in which way the textile products are reinterpreted on the global market? First, there are various drivers that support the entry into the global market. One main driver can be traced back to inspiration. When a textile goes global, inspiration for the brand designers is of particular importance because the textile gets the chance to be perceived all over the world within collections of famous brands. Moreover, the cultural component is also one of the main drivers. Traced back to the roots of the textile which are deeply connected to the culture, going global strengthens the cultural perception. Besides the aspect of perception, going global also helps to better understand different cultures, values, traditions, and the history of societies while also being adapted to global demands. Another significant aspect emphasizes the level of awareness of a textile. Communication about the textile products increase significantly when it becomes global, and the marketing activities intensifies. Consequently, the textile products receive considerable attention and awareness. When considering the approach of going global, the realization of traditional textiles going global can appear in various forms.

In this chapter, we focus on the local production and global trading so that the local aspect remains enormously high. It also addresses the process of going global through outsourcing strategies. A special emphasis is placed on the modernization and adaptation of a tradition to fit global needs, which is included in the process of going global.

The merits of a traditional textile entering the global market can be immense to the local community. Before anything stands its economic valorization. For India, the handloom culture is the second-largest employment provider in its rural areas. It creates employment for over 4 million of its population, and the domestic textiles and apparel market stood at an estimated 100 billion US dollars in the fiscal year of 2019 [17]. To emphasize the importance and relevance of the Indian textile market, export figures can be used, as India is in the third place for its share in the world exports, with 4.2% in 2020. Compared to India, the EU is on the second place concerning share in world exports, with 18.1% in 2020 [33]. On the other hand, when it comes to employment, one can see that during the last recent years, 2018–2019, the employment in the textile manufacturing industry in the EU decreased by 2.3%. [11]. Giving an overview of the whole textile industry, within the second half of 2020, manufacturing goods recorded strong growth in world merchandise trade, especially textiles have a high proportion. When it comes to world merchandise exports of manufactured goods in 2020, textiles had the highest growth with 16% [33].

Traditional textiles, especially with local resourced material, are also highly appreciated in the designer fashion industry. Yves Saint Laurent dedicated his Spring–Summer 1967 collection to Africa: "*He created a series of delicate gowns from a variety of materials, including wooden beads, raffia, straw, and golden thread. At a time when industrial production predominated, it was a way for the couturier to renew with artisanal techniques"* [26]. The artistic director of Christian Dior in 1989, Gianfranco Ferré, fell in love with India and its craft during a trip. He took inspirations from the design of Indian shawls of Kashmir, and reproduced them in light organza fabrics, worked by Indian artisans (Fondazione Gianfranco [12]. These are just to name a few of the examples where designers reinterpreted traditional elements and textiles to the catwalk.

2 Challenges of Entering the Global Market

As explained in previous segments, traditional textile has its roots deeply embedded with the passing of culture and has seen evolution over time to reach its current form. It is also a crystal of emotion and heritage passed onto one generation to the next, and such history can easily be lost through creation with no attachment due to pure demand. Therefore, it is easy to imagine that by expanding into the global market—whether in terms of production or exportation—it will come with its risks.

2.1 Exploitation and Plagiarism

Exploitation is a large keyword to such advances. One embodiment of such would be financial exploitation: where there is a system that intermediary profit, creating an enormous difference for the labor payment and its retail prices. Take for example the case of *"hupil"*, a colonial style garment, sewing together five or more pieces of home woven cloth. Having been widely adopted in the indigenous areas of Mexico, and a vital source of income to women of the community. *"There have been countless accounts of traditional garments, such as the "huipil", being bought from artisans for a fraction of what it actually costs and then deconstructed for accessories to be sold elsewhere. This type of exploitation is unfortunately not unique to the South American country"* [30]. Although in 2021. Zara was also found to have copied the Hupil designs, the company responded to not have intentionally copied or been inspired from them [23].

Exploitation of culture, or in another sense cultural appropriation, is also a discussion to take note of. Cultural appropriation is framed as *"the taking of intellectual property, traditional knowledge, cultural expressions, or artifacts form someone else's culture without permission"* [8]. Commentators also state: *"The very cultural heritage that gives indigenous peoples their identity, now far more than in the past, is under real or potential assault from those who would gather it up, strip away its honored meanings, convert it to a product, and sell it. Each time that happens the heritage itself dies a little, and with it its people"* [14].

A significant example in this context we found is of a small Laotian community called Oma, whose designs were replicated by the luxury fashion brand Max Mara

[32]. The design pattern consists of traditional distinctive craftsmanship with the indigo dye and red embroidery. The Oma women earn income through the sale of their craftwork which helps them toward a better health and education for their families. Despite several complaints and awareness raised in the international community, Max Mara did not acknowledge or compensate the artisans for their work and creativity.

The exploitation can range into various fields due to its nature. The commercial use of cultural or religious motifs could be detrimental to the source of tradition, and tamper with its identity and heritage [8]. One of the main discussions toward these allegations can be that they are lacking compensation toward the origin of the culture and is not contributing to the source community in any way: borrowing without regards. This contains the possibility to dilute the original heritage of the textile, especially in cases of religious or cultural items, by being used for purposes outside their traditional settings.

In a not-so-distant past of India, history evidence how the local textiles from India became a source of sophisticated clothing across Europe and how the local production came to a decline through the hands of the Britain colonialists. In an article by Ritu [19], she highlights that even the everyday clothing needs of the Roman togas and the Greeks were met by India. It was not only the woven clothes from cotton, muslin, or silk, but also a variety of the patterned fabrics were brought to the different parts of the world from India. The printed textiles were referred to as *Chintz*, derived from the word *Cheent* and the Indian Kashmir motif the Paisley. While the colonial regime of the British East India Company profited from trading the Indian textiles in Europe, the local European textile merchants were unhappy. Leading to several bans in France for the import of Chintz between 1686 and 1759, and in Britain between 1700 and 1774 [3]. After the Industrial revolution began, the establishment of cotton and textile manufacturing was observed across the England. These productions were based on the traditional printing, weaving and embroidery techniques, extensively studied by the John Forbes Watson and published in 1866 in the book *The Collections of the Textiles Manufacturers of India* [19]. The British imposed high taxation of 75% for exports from Bengal in India [32] which was a major hub of cotton and muslin production and exports leading to the decline of the fine muslin production and exportation from Bengal.

3 The Case of Nishijin Brocade—Modernizing a Tradition to Fit Global Needs

When people think of Japanese culture, kimono is one of the most iconic items comes to mind. The domestic market of Japanese formal wear, or more simply put kimono wear, has been seeing a steady decline in the past recent decades. The market recording nearly 2 trillion yen in the 1980s is now a 280 billion market in 2020 [35], due to the westernization of daily wear for the Japanese. Kimono is now an occasional item, worn only at weddings, graduations, or garden parties, etc. and the global COVID 19 pandemic has made such occasions even more scarce, resulting in its increased pace of decline. Although kimono holds a strong global presence as an embodiment of Japanese culture, its future in the domestic markets has been far from bright.

Despite this trend, one key player has successfully made its entrance in the global market and has been showing steady growth: the Nishijin brocade (Nishijin-ori in Japanese). The name Nishijin has had several theories in its name origin: but the most well-known to come from the western area of Kyoto, which is birthplace of this fabric, "nishi" meaning west, and "jin" meaning quarters. The birth of the Nishijin brocade dates to the fifth or sixth century, from the imported silk farming and fabric from neighboring countries such as Korea and China. Its characteristic lies in its production method of dying the fabric only after it has been woven. This enables the fabric to be tougher and more resistant to wrinkling, creating a silky sheen to the finished product. The brocade also embeds a certain foiled thread *haku* [18] and twists its threads in complicated and numerous patterns to create a distinctive finish. Nishijin brocade is most seen domestically as the fabric of *obi*, a belt-like garment wrapped around the abdomen on a kimono.

The Nishijin brocade's emergence in the global market was far from success until the recent years. Masao Hosoo, CEO of Hosoo Ltd., home to one of the most prestigious Nishijin brocade houses established in 1688, saw this deterioration of the Kimono industry and found opportunity in this time of crisis. Masao, prior to his heir in the family business, had worked in Itochu, one of Japan's largest trading companies, and experienced product control of a fashion brand based in Milan: and this experience enabled him to realize the true value of Japanese textile and its possibilities. In his eyes, for the Nishijin brocade to enter the global market, the traditional system had to undergo reformations to successfully make way into the new audience, unfamiliar to the textile.

The first reformation was to be able to complete all steps related to manufacturing in-house. Up until that point, it was common for textile makers, especially concerning processes that required specific techniques, to be outsourced according to their individual steps. Masao decided best that the company absorbed all these separate manufacturing steps to the company, and to be able to create the Nishijin brocade (from design to manufacturing) completely in-house. This enabled not only full control of the manufacturing speed and employment, but also aided in preserving the Nishijin technique that was deteriorating due to its lack of successors. It also resulted in accumulating data and technology involving the technique and carried a crucial key to evolving the craft and tradition to modern needs.

The second attempt involved a more direct approach. Masao brought the Nishijin brocade to an international trade fair *Maison et Objets* in Paris in 2006. Despite having no orders or deals, the uniqueness of the said brocade (embedding a specific foil, and the strongly twisted thread) held a special placement amongst its competitors. The participation at the international fair also resulted in Masao choosing a clear target audience for the brocade. He declares his hopes to become the "*Ferrari of woven textile*" [34] and targeted specifically the high-end luxury market. This target is in accordance with Nishjin's origin, where the fabric was largely used for costumes of wealthy aristocrats and samurais. This participation led to the encounter with Peter Marino in 2008, an architect in charge of Christian Dior's interior design at the time. Marino requested a store wallpaper with the Nishijin brocade, and this was the opening chapter for its global, and furthermore, global luxury market emergence.

With the success of going global, came its compromise and consequences. To fit the needs of Marino, the Nishijin house had to undergo another reformation by creating a completely new loom. As stated earlier, the original usage of the textile was for *obi*, a belt-like fabric, and therefore was created in the fabric width ranging from 40 to 50 cm [36]. This new request required a width of at least 150 cm, which was impossible to weave with the pre-existing looms. A fully renewed loom, that would be able to weave not only in varying sizes, but also use delicate fabrics and threads according to the needs had to be implemented. The investment into such enabled great agility to meet specific needs, and widened variety in future collaborations, such as Leica [36], Chanel, Hermes, and Louis Vuitton, and is now a leading traditional textile within the market.

"What the global market requires from Japan, is not the traditional Japanese print Wagara; but instead, a modern textile with the essence of the long history and craft cultivated in Japan" [16].

The success behind Nishijin and its emergence in the global, and more specifically the global luxury market can be distinguished by three key factors. The first being the reformation of its manufacturing structure to fit the needs of the fast-changing industry environment. This is more of a technical standpoint but investing in such adaptation offers textile possibilities in branching into the more modernized fields with agility. Nishijin was fast to make the decision to be lenient in its sizes and patterns and this became crucial factor in the coming years to differentiate themselves from the other kimono-based textiles from Japan.

The second key would be the boldness of having chosen a specific target audience that matches with the textile heritage. By choosing a clear target, it gives a clearer idea of the aesthetics of the textile, without having to explain its background in detail. Since the target was in line with its cultural history, this specificity in the global target did not interfere with the already existing local understanding. On the contrary, it reinforced the high-end luxe image of the textile and was able to brand itself well in both local and global markets.

The last point would be the ability to maintain a good balance between innovation and keeping its original cultural heritage intact. This balance has always been a challenge not only in the textile industry, but also in various cultural fields: how can we reinterpret a traditional, local-oriented culture to fit modern needs? As for the Nishijin brocade, along with its agility to adapt its mindset to the current needs, had in mind the attitude of not only honoring their heritage of over 1200 years, but also thinking of the coming years to preserve the culture for the future generations [25]. The message of its origin being strong enough not to be diluted, became what can be considered a pillar to its balance in innovation and authenticity.

4   The Contradicting Paradigms of Innovation and Preservation

Investigating some examples of traditional textiles entering a global market, it highlights a reoccurring paradox: *How do we balance innovation and the preservation of a traditional textile's authenticity?* From one approach, innovation and reformation of the original textile is inevitable to enter a modern-day global market. With the rapid change in consumer needs and globalization of the overall market, traditional textiles and their manufacturing styles have undergone many adaptations to stay in demand. On a contradicting approach, one may say that such innovations pollute the identity or authenticity of the history a textile carries. Outsourced manufacturing can lead to inconsistent product quality. Marking the origin of the textile has always been a challenge, and by reducing the elements of individuality, could make the tracking even more challenging. Existing literature argues that two existing paradigms have dominated the craft development discourse; the view of craft as authentic and in need of preservation to satisfy the *"global salience for the local ,"* [22] *and the view of craftspeople as "outmoded and in need of modernization"* [24]. We learn that one of the largest difficulties textile products face at the expansion into the global market is the balance of such contradicting ideas, both in its true purpose to preserve the traditional essence of the textile.

5   Conclusion

After analyzing the local environment, as well as the global environment of a textile and the approaches of going global, the key findings should be summarized by referring to the main question: *Which dynamics can make a traditional textile influence a local environment, and then become reinterpreted in a global market?*

To sum up the dynamics which can make a traditional textile influence a local environment, differentiated issues must be highlighted. In this part we have given multiple examples of how the textile linked to a local dimension such as availability of raw material or making processes. Then we've tried to underline the cultural and economic consequences a specific fabric could bring into a local culture. While considering the reinterpretation in a global market, the process of going global often includes the modernization and adaptation of a tradition to fit global needs. These adaptations are often met with challenges with contradicting aspects to preserve its authenticity, while meeting its demands. Throughout our research, we observe that the local textiles products can reach a globalized audience who may appreciate the original patterns and techniques used in the textile materials. But much care is needed, and the companies need to act responsibly in preserving the rights and needs of those who produce them locally without any exploitation in the local and global environment. We learn that the key to a successful integration into the global market is based on the balance of adaptation without tampering into the heritage the textile holds.

References

1. Agulhon M (1982) Fernand Rude. Les Révoltes des Canuts. *Romantisme* 12(38):161–162
2. Allix A (1954) Fernand Rude, C'est nous les canuts. L'insurrection lyonnaise de 1831. Géocarrefour 29(4):340–341
3. Bekhrad J (2020) The floral fabric that was banned. BBCpage. https://www.bbc.com/culture/article/20200420-the-cutesy-fabric-that-was-banned
4. BizVibe (2020, April 22) Global textile industry factsheet 2020: Top 10 largest textile producing countries and top 10 textile exporters in the world. Retrieved October 7, 2021, from https://blog.bizvibe.com/blog/top-10-largest-textile-producing-countries
5. Bhardwaj V, Fairhurst A (2010) Fast fashion: Response to changes in the fashion industry. Int Rev Retail Distrib Consum Res Abberivation 20(1):165–173. Retrieved October 10, 2021, from https://www.researchgate.net/publication/232964904_Fast_fashion_Response_to_changes_in_the_fashion_industry
6. Barnett A (1919) Quelques observations sur le tissage des tissus péruviens. J Soc Am 11(1):135–136
7. Cardon D (1999) À la découverte d'un métier médiéval. La teinture, l'impression et la peinture des tentures et des tissus d'ameublement dans l'Arte della lana (Florence, Bibl. Riccardiana, ms. 2580). Mélanges de l'École française de Rome Moyen Âge 111(1):323–356
8. Cerchia REC, Pozzo B (2021). New Frontiers of Fashion Law. MDPI AG
9. Coatsworth JH (2004) Globalization, growth, and welfare in history. In: Suarez-Orozco M, Qin-Hilliard DB (eds) Globalization: Culture and education in the new Millennium. University of California Press, pp 38–55
10. Dato M (n.d.) Silks for the Crown: five partnerships of merchant manufacturers in eighteenth-century Lyon. Retrieved October 9, 2021, from https://core.ac.uk/display/211237116?recSetID
11. EURATEX (2020) FACTS & KEY FIGURES. OF THE EUROPEAN TEXTILE AND CLOTHING INDUSTRY. Retrieved October 9, 2021, from https://euratex.eu/wp-content/uploads/EURATEX-Facts-Key-Figures-2020-LQ.pdf
12. Ferré FG (2019) Fondazione Gianfranco Ferré. Fondazione Gianfranco Ferré / Gianfranco Ferré / Biography. Retrieved October 8, 2021, from https://www.fondazionegianfrancoferre.com/home/biografia.php?lang=en
13. Gascon R (1952) Structure et géographie d'une maison de marchands de soie à Lyon au XVIe siècle. Rev Geogr Lyon 27(2):145–154
14. Greaves T (1999) Intellectual property rights for indigenous people: A sourcebook. Society for Applied Anthropology.
15. Grand View Research (2021) Textile Market Size Worth $1412.5 Billion By 2028 | CAGR: 4.4%. Retrieved October 07, 2021, from https://www.grandviewresearch.com/press-release/global-textile-market
16. Hosoo M (2021) Nihon no Biishiki de Sekaihatsu ni idomu. Daiyamondosha
17. IBEF (2021) Brand India. IBEF. Retrieved October 8, 2021, from https://www.ibef.org/exports/handloom-industry-india.aspx.
18. KOGEI JAPAN (2010) Nishijin Brocade (Nishijin Ori). KOGEI JAPAN. Retrieved October 5, 2021, from https://kogeijapan.com/locale/en_US/nishijinori/.
19. Kumar R (2021) https://thevoiceoffashion.com/fabric-of-india/opinion/Plagiarism-of-Indian-Design-Has-Gone-on-for-Years-4633
20. Lee J, Kim, J (2010) A study on consumer behavior and preference towards textile materials with environment-friendly treatment. Journal of Fashion 14(3):128–145. Retrieved October 9, 2021, from https://www.researchgate.net/publication/264071114_A_Study_on_Consumer_Behavior_and_Preference_towards_Textile_materials_with_Environment-Friendly_treatment
21. *Les révoltes des Canuts (1831–1834) PDF Francais - PDF FILES* (n.d.) Deedr.Fr. Retrieved October 9, 2021,from https://www.deedr.fr/stucmiacieslith1973/xizehuoixlz-101249/
22. Manzini E, Meroni A (2012) Catalyzing social resources for sustainable changes. Social innovation and community centred design. Product-Service System Design for Sustainability. Sheffield, UK, Greenleaf Publishing

23. Marriott H (2021) Mexico accuses Zara and Anthropologie of cultural appropriation. Retrieved from https://www.theguardian.com/fashion/2021/jun/01/mexico-accuses-zara-and-anthropologie-of-cultural-appropriation
24. Munzer SR, Raustiala K (2009) The uneasy case for intellectual property rights in traditional knowledge. Cardozo Arts & Entertain Law J 27:37–97
25. Miyakoshi Y (2018). Imagination is the strategy!? Interview with 12th Creative of the Nishijin house. Colocal—Learning to eat live locally. Retrieved October 8, 2021, from https://colocal.jp/topics/lifestyle/people/20180123_109622.html
26. Musee Yves Saint Laurent (2019) The Spring-Summer African collection. Musée Yves Saint Laurent Paris. Retrieved October 8, 2021, from https://museeyslparis.com/en/biography/collection-africaine-pe
27. Notten P (2020) Sustainability and circularity in the textile value chain. Global Stocktaking. Retrieved October 9, 2021, from https://www.oneplanetnetwork.org/sites/default/files/unep_sustainability_and_circularity_in_the_textile_value_chain.pdf
28. Russel M (2020, July 24). Textile workers in developing countries and the European fashion industry: Towards sustainability? European Parliament. Retrieved October 9, 2021, from https://www.europarl.europa.eu/thinktank/en/document.html?reference=EPRS_BRI(2020)652025
29. Silk manufactures of Lyons (1868) Scientific American 18(26):406–406
30. Stanton A (2020) The issue of exploitation and appropriation of artisan textiles. Matter Prints. Retrieved October 7, 2021, from https://www.matterprints.com/journal/community/appropriation-of-artisan-textiles/
31. Strapagiel L (2019, April 11) MaxMara allegedly stole designs from a Laos community for these pricey dresses. BuzzFeed News. Retrieved October 14, 2021, from https://www.buzzfeednews.com/article/laurenstrapagiel/maxmara-plagiarism-oma-people-laos
32. Taylor J (1840) A sketch of the topography and statistics of Dacca. G.H. Huttmann, Military Orphan Press, Calcutta, pp 301–307
33. WTO. (2021) World Trade Statistical Review 2021. Retrieved October 7, 2021, from https://www.wto.org/english/res_e/statis_e/wts2021_e/wts2021_e.pdf
34. Yanagihashi Y (2014). Prestige House of "Kyo" Collaborating with Global Top Brands with Nishijin Ori. PROJECT DESIGN MONTHLY ONLINE. Retrieved October 6, 2021, from https://www.projectdesign.jp/201502/haveachance/001868.php
35. Y. P. R. (2021). Kimono Market Research. Market Research and Marketing Analysis by Yano Economic Studies. Retrieved October 6, 2021, from https://www.yano.co.jp/press-release/show/press_id/2441
36. Yamagata M (2019) Japanese creators passing on Elegance to the Future. Hearst Fujigaho, Richesse No.30 Winter Issue pp 257

An Exploratory Study on Sustainable Fashion Approaches in Indian Movie Industry

Veena Rao and **Kartikey Goswami**

Abstract Fashion is an integral part of the movies and plays 'a central role in constructing representations of identities and place' [17] therein. Designing costumes involve a lot of research into understanding the character and the time period they are placed in. For instance, Neeta Lulla, designer for the film 'Jodhaa Akbar,' worked on almost 2600 costumes in the entire film with each costume worn by the protagonist costing an amount of Rs. 2 lakhs. In the period film 'Veer' there were 6 armours designed for Salman Khan, each of which was made at a cost of Rs. 20 lakhs. The question that arises is, what happens to these costumes after the completion of the film? Does it go to the landfill, or is it used in other movies, or is it rented? Do the designers address the environmental concerns in designing and discarding the costumes or is the very concept of sustainable fashion 'an oxymoron' [19]. Qualitative study with focus group discussion was employed involving different stakeholders such as actors, models, costume designers, production managers with an aim to identify the sustainable approaches followed in the Bollywood industry.

1 Introduction

Indian film industry is acclaimed as the world's largest producer of films with a consistent growth curve measured across the number of films produced and distributed in a year [9].

The Deloitte analysis estimates that the direct gross output generated by creative industries (composed of film, television, and online video services industries) is INR 115 k crore and provides direct employment to 8.5 lakh people. The direct and indirect accounting of the movie industry is estimated to generate a gross output of INR 349 k crore and total employment of 26.6 lakhs [10, 42].

V. Rao (✉) · K. Goswami
Department of Design, Manipal School of Architecture and Planning, Manipal Academy of Higher Education, Manipal, India
e-mail: veena.rao@manipal.edu

© The Author(s), under exclusive license to Springer Nature Singapore Pte Ltd. 2022   135
S. S. Muthu (ed.), *Sustainable Approaches in Textiles and Fashion*,
Sustainable Textiles: Production, Processing, Manufacturing & Chemistry,
https://doi.org/10.1007/978-981-19-0874-3_8

With the growth of Indian film industry, the Bollywood productions offer opportunity to global fashion brands for collaborating with designers for film production, red carpet dressing, and product placement [2].

The cross pollination between the film industry and the fashion industry is a unique feature of the Indian film industry. It consists of people working at various levels, which includes independent tailors, designers, stylists, menswear shops, and theatrical supply shops [8, 48]. The association of fashion designers with the film industry has been a development since the 1990s and has grown ever since. Initially the costumes of the film industry were associated with a set of workers known as 'dressmen' [23, 48]. The transition of the 'dressmen' to the association of fashion designers with the Indian film industry has been transitional. Historically the chief source of costumes (ready-to-wear) is associated with dresswalas or dressmen who had costume supply shops with large stock of items for rent (dance costumes for the side dancers) [49]. The lead actors brought their own costumes or wore the costumes worked upon by the darzis who were not necessarily part of the studio. The infiltration of the costume designers started in the sixties wherein the role of the dressmen went beyond the costumes to define the look of the lead actor. Eighties and nineties witnessed the costume designers participating in the fashion shows, opening the boutiques, and bringing in-house tailors [5].

With the rise in fashion markets and media industries in India, the dress designers have become more important as an 'image consultant' who can consolidate a star's positioning in the worlds of both film and fashion. The earliest association of the designers with the film industry is credited to Bhanu Athaiya who won Oscar for her work in the movie 'Gandhi' [39].

The costume designers have now become an integral part of the movies. They have a far more significant role than just dressing of a costume [14]. In the words of Edith Head, Hollywood costume designer and winner of 8 Academy Awards, 'What a costume designer does is a cross between magic and camouflage. We create the illusion of changing the actors into what they are not' [40]. The biggest endorsement of fashion in films is the allocation of a budget and dedicated departments for costumes. It is observed that the average budget spent on making Bollywood movies ranges from INR 20 crore to 50 crores with some movies made for INR 150 crore of which a considerable amount is allocated for the costume department. For instance, the costume budget allocated for the movie 'Heroine' was 1.2 crore in which Kareena Kapoor Khan was seen in 130 outfits designed by the ace designer Manish Malhotra [1].

Similarly, Neeta Lulla, designer for the film 'Jodhaa Akbar,' worked on 2600 costumes in the entire film with each costume worn by the protagonist costing an amount of Rs. 2 lakhs. In the period film 'Veer' there were 6 armours designed for Salman Khan each of which was made at a cost of Rs. 20 lakhs.

The question that arises is, what happens to these costumes after the completion of the film? Does it go to the landfill, or is it used in other movies, or is it rented? Do the designers address the environmental concerns in designing and discarding of the costumes, or is the very concept of sustainable fashion 'an oxymoron?' [19]. In recent years, there has been an increasing awareness of the traditional waste and high levels of emissions of the film industry. One of the most contributing elements is costume production and consumption which is often an overlooked area of research.

Sustainability as a concept has gained attention from various stakeholders involved in the fashion business and academic research, most of which are oriented towards sustainable practices for a circular economy. Similarly, the fashion and its cross pollination with the film industry has been explored in terms of the influence of films on fashion, the character portrayal, fashion film advertisement, etc. The interlinkage of sustainable fashion and the film industry has not been explored in the academia. Through this study, we thus seek to understand the concept of sustainability, and sustainable practices followed in movie industry through literature review and focus group discussion. The findings of the study are summarized by mapping the sustainable practices with different forms of sustainability to provide insight on efforts laid by the industry and highlight the scope for the movie industry to move toward 'Green Cinema.'

2 Literature Review

Costume is an essential element of the film. Throughout the history of the film industry, costume designers have worked on the vision of the director of the film, conceptualizing or replicating clothing based on character in the film or the historical period. While costume and film are integral, the use of sustainable fashion in movies remains ambiguous [16, 30].

This research intended to explore the sustainable fashion approaches in movie industry explored the literature available sub-divided into the following areas:

1. Defining sustainable fashion
2. Sustainable fashion initiatives in Indian movie industry

2.1 Defining Sustainable Fashion

Sustainability has become a topic of research within fashion studies expanding to the areas of environmental studies, urban studies, landscape architecture, etc., accounting for varied interpretation and conceptualization of the term 'Sustainability.' For fashion researchers, the concept of sustainability has been addressed along with the areas of environmental and social sustainability involving the issues of waste generated, reuse, recycling, fair labor practices, etc.

Dissanayake and Weerasinghe [11] identify different elements along the product life cycle from the selection of the raw material to the reuse or recycling which involves.

1. Resource efficiency by the use of renewable and sustainable raw material, reduction of resource consumption, and waste minimization.,

2. Circular design of the products which includes the strategies of design for longevity, design for disassembly, design for composting.,
3. Product life extension through repairing services, sharing platforms for collaborative consumption through sharing, renting, and leasing., and
4. End of circularity through reuse, remanufacturing, and recycling.

The circular design framework as explained by Vecchi [46] takes into consideration all the stakeholders of the fashion industry which includes designers, producers, retailers, and customers through the lifecycle of the garment. It includes.

1. Resources (use of environmentally friendlier materials along with the end of the product's life),
2. Design (design with longevity, design for rebirth by strategies of re-use, repair, redesign, and recycle; minimize waste through smart production techniques etc.,),
3. Production (minimal use of chemicals, consideration for energy and water consumption, use of technology and innovation in production),
4. Retail (establishing take-back model, renting clothes, rehashing clothes, swapping, providing services for repairing, styling etc.,),
5. Consumption (explore options beyond fast fashion), and
6. End-of-life (reuse and upcycling process).

Gordon and Hill [18] suggest a framework of six themes for Eco-fashion which includes.

1. Repurposed and recycled clothing and textiles (emphasis on second-hand clothing and recycling fibers into new textiles),
2. Quality of craftsmanship (considers super craftsmanship for clothing with lasting value),
3. Material origins (explore sustainable fibers),
4. Textile dyeing (explore low-impact or sustainable alternatives for processing),
5. Labor practices (includes health and ethical well-being of the workers in the fashion industry),
6. Treatment of animals (responsible and humane use of animal products).

Fletcher [15], takes into account 4 key phases that play a prominent role in establishing interrelationship between sustainability and the fashion and textile sector which includes:

1. Material diversity: It involves establishing a sustainable way of thinking of fiber alternatives which enable the designers to move from a dependency of a few fibers to a portfolio of fibers with low resource intensity (such as organic cotton, organic wool, peace silk, lyocel, recycled fibers etc.).
2. Ethically made: Fashion and textiles industry is a high-impact sector with complicated industrial chains involving labor, energy, water, and other resources in the conversion and production of textiles and clothing. Addressing these complex issues for sustainable fashion would require improving individual production process that helps minimize energy and resources, on one hand and

ethical practices on another hand that involves developing better relationships with workers, child-labor-free production, etc.

3. Use matters: This phase in the life cycle of the garment requires an approach which focuses on process (improved efficiency and better control of washing), product focus (designing fabrics and garments that cause less impact during laundering), consumer focus (designing products function that influences the habits and values associated with cleaning of clothes).

4. Reuse, recycle, and zero waste: The disposal phase in the life cycle of the garment requires textile waste and management mainly through reuse and recycling. Waste management strategies of reuse, repair, and recycling play a significant role in reducing the environmental impact of the fashion industry. A zero-waste vision for the industry requires the ideas of the cycle where the waste is reconceived as a useful, essential, and valuable component of another product life cycle.

The circular vision in the fashion sector aims at minimizing the production and consumption loop by using ecological and sustainable raw material and recycling and reuse of clothing products for other production [21].

Circular economy is also explained based on three principles: design out waste and pollution, keep products and material in use, regenerate natural systems [26, 47].

The different models assessed lay importance on considering sustainable strategies that can reduce waste, extend the use of the products, discard in a sustainable way (recover the output generated), and ethical practices.

2.2 Sustainable Fashion Initiatives in Indian Movie Industry

The objective of this study is to understand the practices followed by the movie industry with reference to costume production and consumption with a focus on sustainability. The literature on sustainable fashion initiatives is explored for the production of costumes in cinema, celebrity endorsements, and movies based on sustainability to understand the patterns of sustainability that are associated with the cinema.

2.2.1 Costumes in Pre-production, Production, and Post-production

There has been a long association of costume designers with the movies and the success of the movies is emphasized with the role of costume designers associated. The inclusion of awards for the best costume designer at national and international platforms such as Oscar introduced in 1948 emphasizes the role of costume design for the success of a film [6].

The designing of the costumes and the production of the costumes for the movies involve various stages that include designing, making, use, and disposal [24] which can be classified as pre-production, during production, and post-production stages of making a film.

The planning for the designing and production of costumes is based on the budget for the costume design department which determines the production or relying on the clothes that are bought off the rack [12]. Contemporarily, costumes are generally bought from the high street and fast fashion brands depending on the character and the production budget. Fast fashion brands generally use low-paid labor and synthetic material which are believed to be polluting to a greater extent. Even when the costuming department is involved in the designing and development of the costumes, it is found that materials are sourced from suppliers who are involved in similar supply chains. The wastage of the material is also found in the form of spare costumes that are often not used or fabric offcuts during the production of the costumes [7].

The costumes used in the movies with big budget productions are often stored, displayed, or sold as collectable items which is not often seen with low budget production houses. In such production houses the costume designers with limited avenues to discard the costumes often keep and store them, creating more wastage since these are not being reused [7].

The most common sustainable initiatives [31] of costume department in designing for a movie include:

- avoid one-use wardrobe
- use of detergents that are certified
- rent, reuse, borrow, or buy second-hand clothing
- use of sustainable material, and fabric components.

At the end of the project the sustainable options include the following:

- selling costumes to the crew members, to the second-hand stores, or costume rental companies
- storing it for reusing or recycling for other projects
- donating to charity organizations.

2.2.2 Celebrity Endorsements to Sustainable Fashion

The association of celebrities with fashion designers helps in the recognition of designer brands and also to spread awareness of sustainable fashion. In India, several celebrities have endorsed sustainable fashion. A few examples are given below:

Vidya Balan during the promotional activities for the movie 'Shankuntala Devi' has promoted sustainable fashion through the posting of handcrafted sarees and kurtas promoting the local artisans and designers [4].

Priyanka Chopra Jonas dressed up by Roach for the BAFTAs was a gown from the Fall 2020 couture collection of Ronald van der Kemp made using repurposed materials and archival pieces of his previous season [37].

Anushka Sharma, a supporter of the environment, teamed with a foundation called SNEHA to support maternal health. Sharma aims to create awareness about circular fashion by putting her maternity clothes for an online charity sale, the proceeds from the same would be utilized for the cause. She is of opinion that, instead of purchasing new maternity clothes, even if 1% of pregnant women in India bought pre-used maternity clothing we could 'save about as much water as a person drink in over 200 years!!' [20].

Richa Chadda, an advocate of organic lifestyles supports and invests in sustainable labels such as Raw Mango [3].

Gul Panag supports sustainable fashion through a number of social initiatives such as 'Shop For Change Fair Trade' which is a citizen action movement aimed at encouraging poor farmers in different parts of the country such as Telangana, Kutch, and Vidarbha to participate in fair trade [43].

2.2.3 Movies Based on Sustainable Fashion Initiatives

- Indrakshi Pattanaik, best known for her National Award-winning work in the Telugu film 'Mahanati,' completed her studies from Symbiosis Institute of Design and Polimoda International Institute Fashion Design & Marketing. The designer endorses sustainable fashion and in news for a Telugu film titled Mishan Impossible. The film stars actress Taapsee Pannu who would be seen in costumes 'made out of disposable waste or biodegradable materials' [32]. 'The amount of wastage from high fashion and mass production is killing our environment and abusing cheap labor. For instance, the amount of water required to make a pair of jeans is equivalent to what a man consumes throughout the year. Thus, I endorse vintage fashion, reuse of old clothes and handlooms' [34].
- Katrina Kaif for the movie 'Jagga Jasoos' released in 2017 had worn eco-friendly clothing throughout the film purchased from 'The Reformation' in Los Angeles, the firm that is known for sustainable fashion brand that works on the philosophy of upcycling and recycling of the clothes to create stylish, organic, and environmentally friendly clothes [44].

Though fashion is considered an integral part of films, the sustainable initiative in the movie industry is not addressed. The celebrity's association with the sustainable brands and the movies oriented toward designing sustainable costumes is observed as a recent phenomenon. Through this research, the association of sustainability and costume designing in the film industry is explored.

3   Research Methodology

This study is executed following the qualitative study through focus group discussion to gain an in-depth understanding of the sustainable practices followed in the Bollywood industry. The data for the research is collected with semi-structured interviews conducted with the different stakeholders involved in the cinema industry which includes the actors, producers, directors, costume designers, and dress makers. The intention of focus group interview is not to make generalized deduction but to evaluate the current state of sustainability in the context of the film industry and identify research directions for further advancement of the unexplored area. Understanding the domain gap of sustainable approaches followed in the cinema would help in directing the future scope of cinema industry to overcome the challenges of sustainability and move toward 'Green Cinema.'

The interviews are conducted following an interview guide that focused the discussion on how the stakeholders of the movie industry interpret sustainability at pre-production, production, and post-production stages of the film. For the study, the interview is conducted for 30–45 minutes with the participants agreeing to be a part of the study following the explanation of the purpose of the study. The focus group involved are posed with open-ended questions that followed the funnel approach (from general to specific) that enabled engaging the participants positively [29]. The contents of the questions asked are given in the table below:

Opening questions	Have you heard of sustainable fashion?
Introductory questions	Please feel free to respond if you hear the word film and sustainability
Transitions questions	How much budget is allocated to designing costumes in the movies? What is the process of designing costumes for films? Do you have a fashion designer associated with designing for movies or do you rely on the dresswalas?
Key questions	How is the planning of the costumes done through pre-production, production, and postproduction? Do the team of costume designers think of sustainability when designing costumes for different characters of a film? Do you source the materials locally and involve artisans in developing the costumes? Are the decision makers in production aware of the concept of sustainability? Have you worked on any projects that involve designing sustainable costumes? Do you think sustainability is possible in films with regards to costumes? Suggest changes that would help costume design in films to be more sustainable
Ending questions	Suggest changes that would help costume design in films to be more sustainable

The opening questions encouraged all participants to get acquainted with the term sustainable fashion without relating it to their profession and aided in focusing on the factual side of the topic.

The introductory question focused on the topic of the study being introduced to the participant involving them to engage in brainstorming activity around sustainability in relation to their professional experience.

The transition questions are designed slowly shifting the focus to the movies and further to the key questions of the study. The participants are asked general questions on budget and designing of the costumes encouraging them to make connections with their personal experience in the movie industry.

The key questions are directed to the participant's knowledge and experience of being involved in sustainable practices followed in designing and production of sustainable costumes in the movie industry. The participants are encouraged to share a detailed account of the practices followed along with relevant personal experiences.

The ending questions are conclusive questions asked to reflect the opinion of the participant connecting sustainability and movie industry. The discussion at the end is summarized by the interviewer and the participant requested for any relevant information to be added, hence, giving more time to the participants to review the focus group interview conducted.

4   Results and Discussion

4.1   Costume in Pre-production, Production, and Post-production

The designing and production of the costumes in the movie industry are based on the budget allocated to a film. The movies based on the period or historical concept require intricate work which in turn requires huge budget allocation. Movies based on urban lifestyle require less budget allocation.

A period film like 'Bajirao Mastani,' requires the involvement of the costume designer to conduct research in understanding the costumes used in the historical period. For such movies, the costume designer along with a team of tailors involves the karigars to replicate the grandeur of the period which increases the budget allocated to costume design in the movie industry. In contrast, a Rohit Shetty film would require the costume designers to spend time in designing the costumes for the character and does not require intricate work and hence the budget allocated is compromised. However, if a famous actor/star is involved the costumes for the character would be purchased from luxury brands or made-from-scratch wherein the amount allocated for costume design is not compromised. There are four categories of casting in the movie industry which include main cast, secondary cast, tertiary cast, and the background cast. For the main cast the costumes are customized and frequently purchased from luxury brands located in London or Dubai. For the secondary and the tertiary cast the costumes are purchased from the local brands or fast fashion brands such as Zara or H&M. For the background cast, the costumes required are picked up from the streets or rented.

Based on the budget allocated the costume design department involves designing and production of the costumes required for the movie. The department is headed by the costume designer who closely works with the artisans and the tailors in producing the costumes required.

Pre-Production: The design of the costumes is based on the script and the char-acter. The director discusses the script with the costume designer. The costume designer starts working on design and production considering a range of factors such as character, the set design, camera, lighting etc. This brings the difference between a fashion designer and a costume designer. A costume designer needs to design based on the vision and script of the director whereas the fashion designers work based on creativity and trend forecasting. A costume designer needs to work on bringing out the character as envisioned by the director and decide costume produc-tion. For instance, the character of Sonam Kapoor in the movie 'Aisha' is depicted as a modern girl who loves to dress up and does not compromise on the way the character looks. In such cases the costume designing is done by the designer from scratch, keeping in view the international trends and brands. The design of the set also plays a prominent role. The costume designer develops a look board that portrays all the main characters and secondary characters. This is followed with sourcing of the material and stitching of the costumes or purchase of the costumes. The trial or look test is conducted based on which the designs are finalized. This is about the pre-production of the costumes. In a few movies prominent fashion designers who design for the lead actor are involved. An example is the movie 'Heroine' wherein approximately 1.2 crore is spent to design about 130 costumes by Manish Malhotra for Kareena Kapoor.

Production stage: The costume designer develops multiple copies of the costume based on the requirements of the script. For instance, if a shot in a movie involves multiple outdoor settings or a longer duration, the copies of the same costume are kept ready for the lead cast to change. These are also helpful for dressing up the doubles who carry on a few shots for the lead cast.

Post-production stage: After the movie is packed-up, the costumes are stored in the production house in the 'peti' or given to background actors or sold to dresswalas who store in their costume shops and rent these to other production houses. The costumes worn by lead actors are not generally sold to dresswalas. They are sometimes taken back to the designer house (if a fashion designer is involved) or taken back by the lead actors as a memory of the character played. In a few cases they are auctioned, and the auctioned amount is given to the charity. For instance, the costumes worn by Akshay Kumar and Sonakshi Sinha in the movie 'Jocker' were auctioned, and the amount was donated for animal welfare.

4.2   Practice of Sustainability in Movie Industry

The participants of the study voiced the difficulties of bringing sustainability into the production of costumes in the movie industry. They explained sustainability as a concept which needs holistic approach including environmental and social concerns involving different stakeholders such as farmers growing the material, spinning units, weaver, etc. In making the costumes, the costume designers are hired by the producer who is bound by the budget allocated. The importance in designing is to portray the character as per the vision of the director. However, the interviews pointed out a few practices that support sustainability.

At the pre-production stage the costume designer generally sources the material locally for the tertiary actors and the background staff. For the lead cast, care is taken in sourcing the material or the garment which is organic and comfortable.

For most of the period movies the fabric needs to be elaborately ornamented which is generally given to the artisans or karigars who are specialized in a particular art form. Costumes portrayed in films have a huge impact on the customer's choices of fashion and hence a costume designer needs to be credited for reaching out to the artisans and keeping the art form in fashion.

The costume designers rent costumes for the background cast mainly from the dresswalas. Most of the time the tertiary cast are asked to dressup themselves and come to the set if the character involves wearing a simple shirt, trousers, or the kurtis. This helps in reducing the purchase and wastage of the material. Generally, the lead actors' costumes are not repeated and not given to any other cast. However, in a few cases the costumes stored in the petis or availed from dresswalas are reused by mixing and matching in a way that the costumes are not recognizable or relatable to the original film. In a way, they are redesigned completely in a new manner. For example, the interview with Ayesha Khanna [1] identifies one such attempt by the designer wherein the costume worn by Aishwarya Rai in the film Bunty Aur Babli released in 2005 was redesigned and stylized for the background dance in the film Band Baaja Baarrat released in 2010.

At the post-production stage, the costumes that are used by the cast other than the lead cast are stored in petis at the production houses. Most of the time, the lead cast as a part of their contracts are allowed to pick from their movie wardrobe. The costumes are kept in the petis for a certain duration which ends up in the landfill.

4.3 Myth of Sustainability

The participants of the interview express their concern about the fashion industry leading to environmental pollution but also opine that sustainability in the movie industry is not possible since the focus is to portray the character.

Sustainability is expensive which increases the cost of entire production. The costumes after the production are kept in a peti and used for 4 to 5 films after which the costume goes into the landfill. There is no fundamental solution to the environmental pollution created by the use of costumes purchased. The materials that are sustainable are not widely available. Even if a costume designer wants to work with organic materials, it is not available as per the requirements.

4.4   Insights on Sustainable Fashion in Movie Industry

The initiatives discussed in the focus group discussion are mapped with the six forms of sustainable fashion [25] which include the following:

1. Custom-made/Made to order: custom-made/made to order is a sustainable principle that encourages the retailers and customers wherein the costume is designed and produced as per the demand reduces the wastage of the material or stock on the racks [33].
2. Sustainable design techniques/production methods: Different sustainable design techniques include zero-waste cutting, upcycling, design for disassembly, multi-functional/transformational all of which ensure low environmental impact [27]. The sustainable design techniques also can be classified as different strategic tools of designing such as upcycling, recycling, reconstruction, and zero-waste design techniques [25].
3. Fair and ethically made: Ethically made takes into account the workers right in the whole of the supply chain [22]. It includes attempts to protecting human rights, ethically source and produce, and free from human exploitation.
4. Waste management: Waste management is an important concern in the fashion industry which addresses the strategies to reduce the waste during production and recycling of the resources [27].
5. Thrifting/secondhand/vintage buying/charity: Thrift, second-hand and vintage products that are donated, redistributed, and resold for reuse purpose aids in offering the customers the pre-loved clothes at cheaper rates and hence aids in the product life extension of the clothing [22, 25].
6. Collaborative consumption: Collaborative consumption is the new trend in sustainable consumption which aims at addressing the over-consumption and throwaway culture of products that increase the textile waste leading to environmental degradation. Collaborative consumption encourages reuse of goods and reduces new shopping to prevent excessive textile waste. It encourages renting and swapping of fashion products [25, 28].

The focus group discussion points out four forms of sustainable fashion being addressed in the movie industry:

Sustainable design techniques/ Production methods: In the movie industry, it is observed that the costumes stored in *petis* are repurposed into a new style.

The costumes designed and used and transferring them into resources with further value are observed through the attempts by actors such as Tapsee Pannu in collaboration with national award-winning costume designer and stylist Indrakshi Pattanaik.

The definition of circular economy as an economy where the value of products, materials, and resources is maintained in the economy for as long as possible, and the generation of waste is minimized is addressed in the movie industry through reuse of costumes for the side actors, renting the costumes for the dancers, etc.

Fair and ethically made: The thumb rule in sustainable fashion is inclusiveness which supports the artisans and their crafts [25]. In the designing stage it is noted that the costume designers work toward the use of the local material, the involvement of artisans, and the tailors for construction of the garments.

Thrifting/ Secondhand/ Vintage Buying/ Charity: The movie industry works toward sourcing vintage clothing from the branded stores or collaborating to create costumes from scratch. Most of the costumes after production are sold to the dresswalas and in a few instances opened to auction.

Collaborative consumption: The practice of renting costumes from the costume rental houses or reusing the costumes stacked in the godown is a common sustainable practice followed in the movie industry.

The problem of wastage generated from the movie industry is complex which requires a holistic view on the solutions applicable moving toward sustainability.

One such efforts are found in the efforts of costume designer Sinéad Kidao. She has assisted academy award-winning costume designer Jacqueline Durran on various films including Little Women, Beauty and the Beast, Mary Magdalene, etc.

Addressing the sustainability approach to Hollywood, Kidao created the Costume Directory, to make it easier for designers to find suppliers prioritizing sustainability, environmental responsibility, and fair trade [36]. The Costume Directory is an open resource which 'explains the things to consider when choosing a supplier, brand or factory and also provides links to cooperatives, individual artisans and weavers across the world who are sustaining traditional crafts in local environments' [36].

To responses like, 'sustainability is not possible in film costumes,' Kidao has set successful and award-winning examples, declaring, it is possible. The film Little Women (2019) bagged an Oscar nomination and won a Best Costume Design at the BAFTA. The costumes therein were designed by Jacqueline Durran and Kidao assisted her. The March sisters from Little Women were dressed on 'a foundation of sustainable swaps and thriftiness' [35].

The authentic look that was achieved for the project titled Mary Magdalene that Kidao designed for, could only be made possible with firm affirmation in ethical working conditions the designer belives in. Wool and linen used in a project were handwoven in a cooperative in India and cross-stitch embroidery was done by Palestinian women living in a refugee camp in Jordan [35].

For Disney's Beauty and the Beast, the team of Durran and Kidao challenged themselves to make a costume for the character of Belle 'that was head to toe fair-trade, organic and sustainable,' a 'Green Costume.' A red cape from the film was an upcycled, traditionally woven British Jacob's wool from 1970, the look was made with 12 different fabrics that were certified organic and fair-trade [41].

5 Conclusions

Sustainable practices explored in the movie industry highlight that conscious efforts are not laid to achieve sustainability. There are many efforts that can be related to the concepts of sustainability and the industry professionals need to work toward making universal standards. The costume design department of the industry relies mostly on the fast fashion or the garments available in the local market due to constraints of budget and fast casting of extras who try the costume fit only a few days before the start of the production. The sustainability with reference to the movie industry is only limited to the endorsements of celebrities to sustainable fashion or their association with the sustainable fashion brands.

To bring-in sustainability in the movie industry, it requires all players such as cast, producer, director, costume designers, cinematographer to collectively discuss and consciously take decision to include sustainability in the production of the film. After the production of the film, the costumes could be given to the agencies which recycle/ upcycle the costumes. This requires proper planning and execution. The sustainable initiatives need not be limited to the costumes but could extend to catering services, reusable bottles, trash recycling and composting, etc.

Indian filmmakers have long been trying to match the technological prowess of Hollywood, and recent Indian films are examples that they are moving toward achieving it. When efforts could be made to be technologically advanced, why not make similar efforts to be sustainable in not just the costume department, but all other departments as well.

References

1. Ahluwalia S (2018, June 28) Here is what happens to all the costumes you see in bollywood movies. IDiva. https://www.idiva.com/entertainment/bollywood/what-happens-to-the-costumes-used-in-bollywood-movies/17076752
2. Assomull S (2019, August 21) Luxury Brands Need Better Strategies for Bollywood. Here's Why: Global brands are hungry for a bigger slice of India's $30 billion luxury market but they aren't exploiting the country's colossal film industry to its full potential. Business of Fashion. https://www.businessoffashion.com/articles/luxury/luxury-brands-need-better-strategies-for-bollywood-heres-why
3. Bolly Celebs Who Support Sustainable Fashion (2018, August 22) India.com. https://www.india.com/photos/entertainment/bolly-celebs-who-support-sustainable-fashion-139176/-richa-chadha-139185/
4. Bollywood and its take on sustainability (n.d.) https://www.unnatisilks.com/blog/bollywood-and-its-take-on-sustainable-fashion/
5. Borah J (2018, August 21) Tailors of Bollywood. The Voice of Fashion. https://thevoiceoffashion.com/centrestage/features/tailors-of-bollywood--1194
6. Bug P, Niemann CL, Welle L (2020) Cinema films influencing fashion. In: Bug P (ed) Fashion and film moving images and consumer behaviour. Springer Nature Singapore Pte Ltd, pp 9–28. https://dokumen.pub/qdownload/fashion-and-film-moving-imagesand-consumer-behavior-1st-ed-2020-978-981-13-9541-3-978-981-13-9542-0.html
7. Cama (n.d.) The green movement in the costume department. https://cama.co.uk/costume-recycling-why-is-it-important/
8. Chatterjee D, Vasek C (2020) Bollywood: Cross pollination between film costumes and fashion. Fash Pract 12(2):219–244
9. Dastidar SG, Elliot C (2019) Circular economy and waste in the fashion industry. Laws 8(26):1–13
10. Deloitte Touche Tohmatsu India LLP (2020, May) *Economic impact of the film, television, and online video services industry in India, 2019.* https://www.mpa-apac.org/wp-content/uploads/2020/07/20200708_India-ECR-2019_Finalized.pdf

11. Dissanayake DGK, Weerasinghe D (2021) Towards circular economy in fashion: Review of strategies, barriers and enablers. Circular Economy and Sustainability, 2021.
12. Enzinger K (2017) Thinking through value transformations of movie costumes. Journal of Extreme Anthropology 1(2):71–74. https://doi.org/10.5617/jea.4889
13. Ferdinand D, Peter D, Eva H (2020) Circular business models in textiles and apparel sector in Slovakia. Cent Eur Bus Rev 9(1):1–19
14. Figueiredo CM, Cabral A (2019) Costumes and characters' construction in cinematographic fiction and drama. In: Montagna G, Carvalho C (eds.) Textiles, identity and innovation: Design the future. CRC Press, pp 73–80. https://www.google.co.in/books/edition/Textiles_Identity_and_Innovation_Design/31BtDwAAQBAJ?hl=en&gbpv=1
15. Fletcher K (2010) Slow fashion: An invitation for change system. Fash Pract 2(2):259–266
16. Friedland NE (2019) Costume and Fashion. https://www.oxfordbibliographies.com/view/document/obo-9780199791286/obo-9780199791286-0020.xml
17. Geczy A, Karaminas V (2016) Fashion's double: Representations of Fashion in painting, photography and film. Bloomsbury Academic. https://www.google.co.in/books/edition/Fashion_s_Double/QLbPCgAAQBAJ?hl=en&gbpv=1&dq=Fashion+plays+a+pivotal+role+in+constituting+a+film+character&pg=PA52&printsec=frontcover
18. Gordon JF, Hill C (2015) Sustainable fashion: Past, present, and future. [electronic resource] London, UK, Bloomsbury
19. Henninger CE, Alevizou PJ, Oates CJ (2016) What is sustainable fashion? J Fash Mark Manag 20(4):400–416
20. Hiro S (2021, June 21) Exclusive! Anushka Sharma raises awareness about circular fashion by putting up her favourite maternity clothes for sale. E Times. https://www.idiva.com/fashion/celebrity-style/bollywood-stars-eco-fashion/18021620
21. Jacometti V (2019) Circular economy and waste in the fashion industry. Laws 8(26):1–13
22. Jestratijevic & Rudd (2018) https://lupinepublishers.com/fashion-technology-textile-engineering/pdf/LTTFD.MS.ID.000145.pdf)
23. Journey of Costume Designers In Bollywood (2012, May 27) Textile Value Chain. https://textilevaluechain.in/in-depth-analysis/articles/textile-articles/journey-of-costume-designers-in-bollywood/
24. Katria S, Shekri S (2012) Birth of Costume in Indian Cinematic World. Fibre2Fashion. https://static.fibre2fashion.com/ArticleResources/PdfFiles/66/6503.pdf
25. Khandual, Asimananda, and Swikruti Pradhan (2019) Fashion brands and consumers approach towards sustainable fashion. In: Senthilkannan Muthu S (ed) Fast fashion, fashion brands and sustainable consumption. Singapore: Springer Singapore, pp 37–54. https://doi.org/10.1007/978-981-13-1268-7_3.
26. Kirchherr J, Reike D, Hekkert M (2017) Conceptualizing the circular economy: An analysis of 114 definitions. Resour Conserv Recycl 127:221–232
27. Kälkäjä R (2016) Sustainable design strategies Examination of aesthetics and function in zero waste and upcycling. Master's Thesis, University of Lapland, Rovaniemi, Finland. https://core.ac.uk/download/pdf/44346785.pdf
28. Lang C, Armstrong CMJ (2018) Collaborative consumption: The influence of fashion leadership, need for uniqueness, and materialism on female consumers' adoption of clothing renting and swapping. Sustainable Production and Consumption. 13:37–47
29. Lewis M (1995, December 4) Focus group interviews in qualitative research: a review of the literature. Action Research Electronic Reader. http://www.aral.com.au/arow/rlewis.html
30. Linden J (n.d.) Conceptualizing the fashion film: The issue of sustainability in fashion and the fashion film festival (Master's Thesis, University of Amsterdam, Netherlands). https://www.academia.edu/36965048/Conceptualizing_the_Fashion_Film_The_Issue_of_Sustainability_in_Fashion_and_the_Fashion_Film_Festival
31. Lopera-Mármol M, Jiménez-Morales M (2021) Green shooting: Media sustainability, a new trend. Sustainability 13:3001
32. Mazumder, S. (2021, July 7). Taapsee Pannu is making sustainable fashion choices in her next film 'Mishan Impossible'. IDiva. https://www.idiva.com/fashion/celebrity-style/taapsee-pannu-in-sustainable-clothes-for-mishan-impossible/18021901

33. O'Driscoll J (2020, February 28) Is made to order the future of sustainable fashion? Eco-Age. https://eco-age.com/resources/made-order-future-sustainable-fashion/
34. Odisha's Indrakshi behind Mahanati's National Award-winning costumes. (2019, August 24). The New Indian Express. https://www.newindianexpress.com/states/odisha/2019/aug/24/odishas-indrakshi-behind-mahanatis-national-award-winning-costumes-2023707.html
35. Parsons, S. (2020, February 7) Sustainability stories: Sourcing sustainable costume with the costume directory's Sinéad Kidao. Eco-Age. https://eco-age.com/resources/sourcing-sustainable-costume-with-costume-directory-sinead-kidao/
36. Production Handbook: Costume Directory (n.d.) We Are Albert. https://wearealbert.org/production-handbook/costumes/
37. Raniwala P (2021, May 4) Sustainability on the red carpet? Here are the celebrities championing it. Vogue. https://www.vogue.in/fashion/content/sustainability-on-the-red-carpet-here-are-the-celebrities-championing-it
38. Schmitz, S. (2017, July 19). Two costume designers talk sustainable fashion. Vilda. http://www.vildamagazine.com/2017/07/two-costume-designers-talk-sustainable-fashion/
39. Sharda NL, Jaglan V (2012) Costume design in the Indian Hindi Films—A Changing Scenario. https://www.fibre2fashion.com/industry-article/6378/costume-design-in-the-indian-hindi-films
40. Sigler LA (2021) Clothes make the character: The role of wardrobe in early motion pictures. McFarland & Company, Inc., Publishers. https://www.google.co.in/books/edition/Clothes_Make_the_Character/NgIaEAAAQBAJ?hl=en&gbpv=1&dq=Fashion+is+an+integral+part+of+the+movies+used+to+provide+necessary+to+the+film+and+character.&pg=PA23&printsec=frontcover
41. Sinead Kidao [@thecostumedirectory] (2017, March 17) @beautyandthebeast is out today! I was an assistant designer to Jacqueline Durran on the job, which had a costume team [Photograph]. Instagram. https://www.instagram.com/p/BRvt8LLhlET/?utm_medium=copy_link
42. Statista Research Department (2021, Oct 18). Film Industry in India - statistics & facts. https://www.statista.com/topics/2140/film-industry-in-india/
43. Sustainable Fashion on the Indian Roll, (2021, Oct 4) Fashinza. https://fashinza.com/textile/fashion-industry/fashion-news-today-sustainable-fashion-on-the-indian-roll/
44. Sustainable clothes for Katrina in "Jagga Jasoos" (2016, Oct 28) Business-Standard. https://www.business-standard.com/article/pti-stories/sustainable-clothes-for-katrina-in-jagga-jasoos-116102800259_1.html
45. The Costume Directory (n.d.). Sinead Kidao. https://www.sineadkidao.com/the-costume-directory
46. Vecchi A (2020) The circular fashion framework—The implementation of the circular economy by the fashion industry. Current Trends in Fashion Technology & Textile Engineering 6(2):31–35
47. What is a circular economy? (n.d.) Ellen Macarthur Foundation. https://ellenmacarthurfoundation.org/topics/circular-economy-introduction/overview
48. Wilkinson-Weber CM (2006) The dressman's line: Transforming the work of costumers in popular Hindi film. Anthropol Q 79(4):581–608
49. Wilkinson-Weber CM (2010) From commodity to costume: Productive consumption in the making of Bollywood film looks. J Mater Cult 15(1):3–29

Revitalising Crafts for Sustainable Fashion

Neha Mulchandani

Abstract With the growing awareness for sustainable development, transition to sustainable fashion can play a vital role for longevity, cleaner environment and well-being of individuals, communities and societies at large. There are many ways to approach sustainability in fashion and linking crafts is one such way of achieving it. The crafts have had strong relationship with sustainability and recently there is revival of interest on traditional craft practices. The local crafts, local knowledge and linking them, their nature and philosophies to modern lifestyle can offer sustainable pathways. India's rich craft traditions form an integral part of the country. They are the reflection of rural community life, the flora, fauna and other elements inherent of the region or geographical area. When it comes to urban settings, the traditional crafts are not much favoured over technology driven products. The chapter discusses the importance of sustainable fashion and focuses on the interweaving of crafts with contemporary design contributing to sustainability. The chapter contributes to the understanding of craft and sustainability relationship and the potential approaches how craft can contribute to sustenance. The chapter also discusses the case studies of how various traditional crafts have been integrated or hybridised to modern design thereby meeting the demands of the modern society.

Keywords Crafts · Sustainability · Fashion · Contemporisation · Co-creation · Transition · Co-design · Niche innovation

1 Introduction

In India, the craft is perceived as a social or rural activity and has been positioned within cultural fields. Despite being the second largest employment providing sector, it has not gained an economic status. Most of the crafts are not connected with modern

N. Mulchandani (✉)
Department of Textile and Fashion Technology, College of Home Science,
Nirmala Niketan, Mumbai, India
e-mail: nehamulchandani@nnchsc.edu.in

© The Author(s), under exclusive license to Springer Nature Singapore Pte Ltd. 2022   151
S. S. Muthu (ed.), *Sustainable Approaches in Textiles and Fashion*,
Sustainable Textiles: Production, Processing, Manufacturing & Chemistry,
https://doi.org/10.1007/978-981-19-0874-3_9

society thus creating huge vacuum between indigenous culture and modern population. They are often considered old fashioned and are undervalued in comparison to mechanised products. Consequently, the craft is not able to compete with such products and slowly declines. The craftspeople, therefore, get affected as they are unable to make a decent living leaving many craft makers looking for different occupation thus causing a break with traditions in the community [38].

These crafts and their practices form part of cultural heritage and thereby should be preserved and revitalised. Not only that, they invigorate local economic development [27]. Because of its shared values and social capital, the craft sector is portrayed as an ethical and socially responsible trade [33]. The sector gives opportunities to generate income and is described through entrepreneurship and artisanal production [8, 24, 29].

The most challenging issue of the twenty-first century is to ensure that present and future ecosystems sustain [34]. The consumers are now slowly becoming engaged to sustainability and as per the survey conducted by Mckinsey survey 2021, 2/3rd of consumers surveyed supported the promotion of responsibilities towards society and environment [14]. As per the study, the young consumers are willing to change and becoming more open to experiment with smaller or lesser-known brands during the COVID crisis. This could serve as an opportunity for crafts sector to position themselves to emerging and new markets. The recent growing demand for sustainable fashion further strengthens opportunities for craft sector to supply the new markets. The recent times have triggered interest in sustainable practices but at the same time consumers would like to follow the fashion trends. The fashion industry must advance today towards sustainable fashion with the new rules set by the user within today's environment. The local crafts have the potential to respond to this trend of sustainable fashion if sufficient support and direction is provided to them.

The chapter stresses how traditional crafts can offer possible sustainable solutions and how craft can play a key role for sustainable fashion. It draws attention to the necessity of learning from traditional skills and the value of craftsmanship for long-term sustainable fashion. The chapter further presents traditional crafts as a viable strategy for sustainability by repositioning the craft knowledge and skills to current lives, thus making it an economic viable sector and at the same time being pillars of cultural continuity and cohesive societies. The chapter also focusses the useful strategies and approaches that previous research have indicated for promoting sustainable development across the context of traditional crafts.

Following this, the chapter discusses the case studies of ethical sustainable brands collaborating with weavers and craftsperson to create new products with balance of tradition and modernity. The case studies discussed will further enable us to understand the practices followed and experiential knowledge which will help further to create new designs and new products.

2 Craft as a Pathway for Sustainable Fashion

The concept of sustainable development was first introduced by World Commission on Environment and Development (WCED) in 1987 in the report "Our Common Future" commonly called as the Brundtland Report [36]. For the very first time, the concept of sustainable development was outlined in the report by quoting it as "development that meets the needs of the present without compromising the ability of future generations to meet their own needs". The concept of inclusive growth which ensures well-being for present population and for future generations forms the central core towards sustainable development. With time, the concept and definition of sustainable development has evolved and the framework calls for collective action by all countries, poor, rich and middle-income to promote prosperity while protecting the planet. The United Nations 2030 Sustainable Development Goals (SDGs) acknowledges the world shortcomings of sustainable key goals for people, planet, peace and prosperity. Sustainable development cannot be achieved till we eradicate poverty in all forms and from everywhere, and by promoting inclusive economic growth, employment and addressing a range of social needs including education, health and social protection while addressing climate change and environmental protection. Though the SDGs are not legally binding, it is necessary that all the governments set the priorities and establish national frameworks for the achievement of the 17 Goals [31].

The sustainable development discussion begins with three pillars-economic growth, environmental stewardship, and social inclusion with the goal of improving people's well-being and nature [3].

The way forward is to encourage changes in different sectors as the world sees some major threats in the future. One of the most important sectors is the fashion and apparel sector. The fashion and apparel industry is the most polluting industry in the world causing a significant environmental impact and therefore this industry is expected to be more responsible for sustainable actions. The industry is largely driven by fast fashion model where revenues are based on selling more low-priced fashion products leading to greater consumption. The so to say fast fashion gives consumers an opportunity to buy cheap products but lack emotional or symbolic value and are cared less. This leads to tons of waste generation [11]. To address the environmental concerns and to incorporate a sustainable approach, Fletcher propagated slow fashion which put simply, is just the opposite of fast fashion. It's based on making ethical and conscious choices encouraging both the consumers and producers to focus on sustainability. It's a movement that advocates discovering sustainable alternatives at various levels such as design, production, consumption, use and reuse. It encourages designers to ensure product quality, inspires them to explore alternative materials to create awareness and responsibility towards fashion production and consumption.

Craft has close connections with sustainable development. Craft is a trade and profession, a product or artefact, a culture, a creative, intentional process of skill and craftsmanship, techniques and materials, and a complex interplay of these aspects. Traditional crafts involve high degree of craftsmanship that draw on cultural shapes, models and designs using inherited techniques [33]. Craft is defined by specialised knowledge, localisation, ethics and authenticity, and the continuation of tradition. It reflects the artisans way of life and the heritage and the skills which are acquired from the past generations. The artisan with knowledge and creativity expresses his individuality and culture through his craft. Craft by their very nature is mostly manufactured as a one-off or in a small quantity, the design of which might not necessarily be culturally rooted in the country of production. The very first examples of the craft were the objects or products for personal use to fulfil basic needs. However, it has now grown to cater to a wider audience, and it has thus become a significant source of income for artisans at the grassroots level [9]. It has high degree of hand input and can be sold to make profits [7]. The crafts of India depict its rich heritage history and diverse culture and the legacy is passed on from one generation to next generation. It serves as powerful tool to sustain cultural heritage and can act as bridge between the modern and traditional. It is also considered to comply with the principles of sustainability with regard to materials and production, the development of knowledge and skills, social responsibility and continuity of cultural and aesthetic experience [7, 33]. The earliest steps towards mastering the human environment are craft and design [34].

At an individual level, craft cultivates skills, knowledge, and values in creators, primarily through reflection on doing. At a cultural level, it acts as a medium for communication where handmade items communicate symbolic language: cultural symbols, forms, motifs, and colours that engage the maker, user, producer, and consumer. At the social level, craft can educate about materials and techniques to innovate in the wider context of sustainability [34].

Craft becomes sustainable through the quality provided by hand/body, skill and practice, materials and design and technique, linking all the disciplines to function and aesthetics presented in Fig. 1.

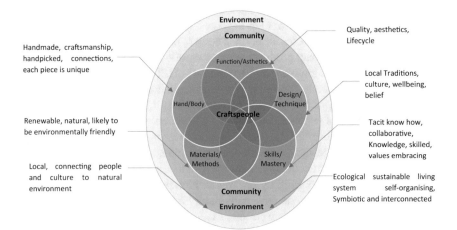

Fig. 1 Craft ecosystem

Hand/Body: Craft involves labour done by hand or with tools held by hand or a handmade product. It is a process that involves head, hand and heart connection. It is anything made by hands connecting to human values from its experimental and emotional knowledge [32]. The craft is the output of mind and an expression of makers love, emotions, thoughts and feelings. According to Fuchs et al. [12] the handmade objects contain creator's "love" and as stated by Job et al. [17] they are said to have "positive social traces" respectively.

Materials/Methods: This deals with the materials and methods of making. Traditional crafts have a deep understanding of nature and use of natural materials. Yair [37] highlights the role of materials in pursuing sustainability. Textile crafts such as hand weaving, knitting and natural dyeing, etc. encourage ecologically conscious production thereby accentuating the imperative of sustainability and offering solutions to climate change, pollution, and waste [28]. The ecological attributes through the use of eco-friendly resources, recyclable material, respect of nature, etc. are not just superficially embodied but are reflected in the culture of sustainable craft as a whole. Some of the traditional textile crafts make use of traditional technologies and are dyed with natural dyes.

Skill/Mastery: Crafts refer to skills and mastery of materials and techniques. Skills refer to technical proficiency and manual dexterity and mastery indicates performance at 'their highest level'. The quality of the crafted product depends on the expertise and skill of the craftsperson.

The perception and skill performance develop with time from the novice to the expert stage, decision making shifts from analytical to intuitive and commitment shifts from detached to engaged [32]. Learning the traditional skills opens the path for a sustainable way of living for the future and applying them in contemporary art constitute an influential method when striving for cultural sustainability.

Design/Technique: The craft objects tend to last much longer and the traditional designs leave deep and long-lasting impact giving an impression of timeliness [23].

Function/Aesthetics: Functional skills involves knowledge and material awareness in order to make high-quality durable crafts. Aesthetic skills refers to crafting beautiful products inspired by nature and surroundings [15]. Craft is closely associated with multiple functions which include social values, cultural identity and aesthetics. The way the materials, which the craft is made of, interact like the fineness of silk, the roughness of linen or the softness and the scent of wool, which act as physiological dimensions can take us back to old memories and experiences. The colours, shapes, material etc. which act as psychological dimensions can remind of culture and traditions [35].

3   Transition to Sustainable Fashion

The necessity of sustainable development has led to the concept of sustainable fashion. Achieving such transitions is possible through the shift from established socio-technical systems to more sustainable modes of production and consumption. Such transitions to sustainability involve long-term, multi-dimensional and fundamental transformations [20].

Many frameworks have emerged for understanding transitions. One such notable framework is the multi-level perspective (MLP) which is employed to conceptualise, understand and promote the transition towards sustainability. Transitions, according to the MLP, occur as a result of interactions within and between three analytical levels: niches, regimes, and a socio-technical landscape. Niches form the microlevel in MLP, which are the spaces where new ideas develop. The niches are where the transitions begin. They are unstable at the beginning with low performance. The innovations are commonly conceived, developed and promoted by small networks of dedicated actors, either new entrants or fringe actors [13]. The socio-technical regime forms the meso-level which is like a deep structure of the existing socio-technical system. It refers to the set of rules and institutions such as shared meanings heuristics, rules of thumb, routines and social norms that run the dominant system and related technical and material elements. It consists of a network of actors and social groups whose actions and perceptions are shaped by these formal and informal rules and institutions. It accounts for the stability of the incumbent socio-technical system which includes the existing technology, industries, supply chains, consumption patterns, policies and infrastructures. Transformations and reconfigurations of regimes are uncommon as they have considerable inertia making it difficult for the innovations to establish [13]. The socio-technical landscape forms the macrolevel which indirectly influence the niche and regime dynamics. It refers to exogenous trends and events such as political, demographic, socio-cultural, environmental and external factors. Even though the changes take place slowly in decades at this level but these changes cannot be changed easily in the short run. The changes at this level create pressure on regime and open a window of opportunities for niches.

The socio-technical transformations do not happen by themselves. According to the MLP, these happen as a result of trajectories and current processes aligning within and across these three analytical levels. Specifically, the developments result when niche inventions progressively gain internal momentum (e.g. through learning processes, improving technologies, opening up markets, finding customers, attracting investment, lobbying policymakers for support and developing social networks). Changes at the landscape level create pressure on the regime and disrupt its stability due to which the niche innovations break out from the micro-level and start diffusing and aligning within the socio-technical system leading to substantial transformation and disruption of the existing regime processes. The alignment of various processes at different levels is important for the transition to take place.

Using this framework of MLP, the fast-fashion concept centres on the socio-technical system of fashion. All the players in the production, distribution and consumption of clothes such as suppliers, fabric and apparel manufacturers, universities, research centres, government regulations, consumer practices, energy infrastructure, water infrastructure, cultural and symbolic meanings, distribution, supply chain and technology make up the socio-technical system of fashion. The elements of the socio-technical system interact with one another and also with other systems such as agriculture or chemistry. The socio-technical fashion regime consists of beliefs, routines, norms and standards of doing something, with the fast-fashion model in the mainstream. It means that the phenomenon is to manufacture/sell/buy cheap clothes that follow the latest fashion trends [4]. This fashion phenomenon which is based on production, consumption, change and waste is fast fashion. Despite the constant changes in trends, the market still experiences the growing speed of fast fashion trends. Fast fashion has dramatically increased the consumption patterns and has caused various environmental problems. The fast fashion business model promotes cheap, fashionable and low-quality clothes and the consumption is based on buy, use and discard. For transitioning to an eco-friendly landscape especially in the fashion industry, new concepts have come to the forefront such as slow fashion, sustainable fashion and ethical fashion as presented in Fig. 2. All of these concepts have similarities and all of them focus on improving matters in terms of environment or social or ethical impact. They seek to mitigate the fashion lifecycle, slow down production, reduce consumerism, use eco-friendly processes by giving more preference to the environment. These new and emerging concepts can be seen as an alternative to fast fashion, which is facing growing criticism for its disastrous impact on the environment [18, 22].

In the niches, are sustainable brands, small businesses or non-governmental organisations with sustainability at their core, that aim to break the regime barriers and offer novel products, cutting edge fashion and processes and promote sustainable practices, small scale production, artisanal work [5].

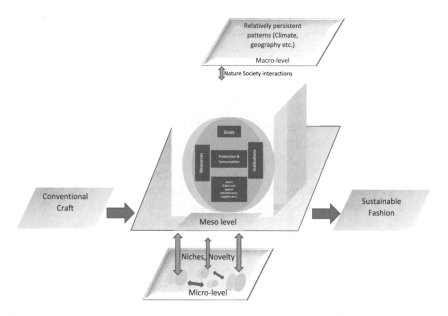

Fig. 2 A framework for research in sustainability science. Adapted from Clark and Harley [6]

4   Contemporisation of Craft as Niche Innovation

To become sustainable and to minimise the growing problem of wasteful consumption, fashion has to explore and set new aspirations. An emotional connection that the consumer may associate with the designed garment or fashion product is needed besides function. This connection as per Fletcher [11] can only be established by producing fashion that is founded on values and skills, as well as being conscious and long-lasting. This change in the pattern of consumption places more relevance to cotemporisation of craft. It can not only pave way ahead for sustainability but also maintain communities and safeguard the environment. The sustenance of craft culture is important for local businesses and can create pathways to stimulate creativity and innovation in the quest for inclusive, equitable, and sustainable development.

These culturally connected traditional crafts have a strong position to respond to sustainable concerns but efforts are needed to bring them to the mainstream market. The craft practices do not remain static but have the capability of evolving, changing and adapting to the needs of modern society.

4.1   Co-creation as an Approach to the Development of Local Craft

Co-creation, an act of collective creativity between designers and artisans, is an approach to the development of craft. The possible synergy between designers and artisan groups can promote the growth and preservation of a local craft in a sustainable and commercially viable manner. This collaboration can likely discover new potential for local crafts [2, 21]. The challenge for the artisans is to produce fashion that caters to the changing tastes of the consumers and market them. It is possible to bridge this gap if the designers assist the artisan communities and improve their processes, achieve a new level of aesthetic excellence and familiarise them with market opportunities. Rubens, 2010 highlights how collaboration maximises the skill and knowledge base that each party provides to the innovation process [26]. The designers direct participation and engagement in a local context by interacting and co-creating with the artisan community help them to understand the problems and shortcomings and address them effectively. Designers play an important role in the design innovation process as they tackle the challenges that the individual or artisan cannot do alone. The initial phase which is ambiguous and intangible referred to as fuzzy front-end phase by researchers involves deep design thinking and collective inputs. It is an exploration phase wherein the main problems are identified, the opportunities are discovered and design approach is formulated based on existing solutions, new ideas and design concepts. This is the critical phase setting the stage for further phases and therefore involvement and participation of all teams is important. The next phase is the design development phase where the ideas generated are developed into concepts, prototypes, and then refined into resulting products.

Figure 3 illustrates the co-creation process and how designers along with the artisans could identify the hidden new opportunities and new ideas.

Also, the design craft collaboration enables sharing of knowledge and skills through experiences thus enriching each other. Both the artisans and designers in this way can capture each other's expertise to make novel and appropriate sustainable products.

Dholakia and Parmar [10] explored to give the craft of beadwork a new meaning and diversify it through the co-creation process. This craft is typically been used for making accessories and has not been explored much beyond that. Not much contemporised pieces are seen due to which the craft is slowly getting pushed towards extinction. The researchers co-created and facilitated the application of craft on apparel by understanding the local setting, craft, skills, products and materials. A thorough study of the craft by the researchers/designers gave them an understanding of the major gap, the craft was facing. This initial phase called as the fuzzy front end of the collaboration process helped the designers to propose practical solutions.

The tacit knowledge can be effectively understood through face-to-face communication [1]. The interactions and discussions with artisans practising the craft for more than 20 years facilitated the researchers to develop an understanding of the challenges faced by the artisans. The craftsperson was accustomed to the traditional typical colour palette, typical geometric designs and made typical products and took long time to create motifs. During the design and development phase, the designers gave the design direction whereas selection of suitable motifs, method and technique were up to the artisans choice. Using visuals in the form of sketches and prototypes during the design development stage helped with communication and artisan-designer interaction. The researchers/designers co-created the contemporised market-oriented collection using the bead craft. The work in the study brings attention to the fact that through the co-creation process artisans were able to discover new creative possibilities as they became aware of the design potentials in beadwork, while the designers could materialise design concepts through a better understanding of craft materials and techniques (Fig. 4).

Realising the importance of Indian heritage, textile designer-maker-artist Swati Kalsi re-interpreted the indigenous Sujni embroidery along with the artisans of fifteen

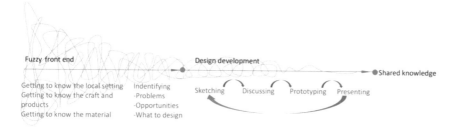

Fig. 3 The craft-design collaboration process proposed by Tung [30]. Reprinted under Creative Commons license

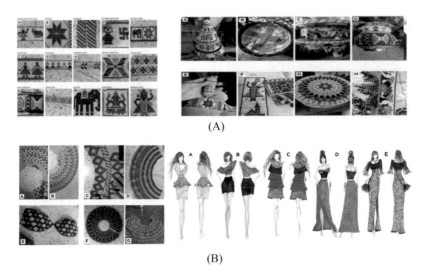

(A)

(B)

Fig. 4 Contemporisation—An outcome of co-creation process. (A) Traditional motifs, colours and products; (B) New motifs, color palette and products. Reprinted under Creative Commons license, Dholkia and Parmar [10]

villages of Bhusura district in the Gaighat block of Muzaffarpur in Bihar. She successfully presented collection at Lakme Fashion week in 2014 leveraging the potential of Sujni embroidery based on design approach which was conceptual in nature along with the understanding of the intricacies of the technique. The designer with an inherent understanding realises that in order to be ecologically sustainable, it's important to develop a critical interface that will extend the traditional knowledge into the vernacular of contemporary design by repositioning itself with the mainstream fashion. The concepts and ideas for the collection originated by engaging in discussion with the embroiderer and the sketches further gave the specific design direction which included the size and balance of the composition, colour, quality and direction of the stitches which further facilitated the designer artisan integration. The designer followed an inclusive process where the artisan was given the freedom to interpret her concepts in their own way. However, the final decision on the aesthetics was taken by the designer. The resulting products were original and new and exhibited the contemporised form which were developed with the objective to increase visibility of craft in urban market [16] (Fig. 5).

4.2 Brand Connecting with the Artisans

Vraj: Bhoomi, is a fashion label that provides ethical fashion by working closely with the artisans of Kutch. The label transforms traditional hand block printing skills to plush fashion whilst maintaining the traditional design philosophy. The brand

Fig. 5 Swati Kalsi, Sujani collection [41, 42]

is engaged in reviving the age-old technique of Ajrakh and revamps it to a new form. Ajrakh is a traditional local textile craft practised by artisans of Kutch using indigenous materials to make functional and decorative textile items. It's an age-old craft and the origins date back to Indo-Egyptian cloth trade. The fabric is created by an intensive 16 step process of washing, dyeing, printing, and drying that requires a high level of expertise and the important part is to keep the colours fast and even. It's a long painstaking process where the artisan prints one colour of a design and washes the cloth till he gets the colour right before printing the next colour [25]. The craft utilises locally cultivated indigo, pomegranate seeds, gum, harda powder, wood, flower of kachika and dhavadi and alizarine for colouring the fabric. Typical motifs printed on Ajrakh are complex geometrical motifs which create starry constellation in indigo, madder, black and white across a length of cloth forming patterns such as Islamic jali windows and trefoil arches. With changing trends, the production of this fabric has lessened to a great extent and is proving to be a threat to the age-old traditions of this textile art. This brands focusses on preserving this craft by reviving it in different ways. The brand collaborates with artisans by giving numerous inputs such as design, quality control, raw material access and manufacturing coordination for the commercialisation process. The brand creates beautiful ensembles using their signature block designs which are contemporary and universal in nature. They retain the traditional colour palette whereas use contemporary motifs blending them in their unique way. The brand also takes inspiration from other art craft forms such as Turkish tiles the pottery and suzanis of Uzbekistan to the traditional Pichwais of Nathdwara. The label tries to translate the cultural heritage into various types of products ranging from clothing with aesthetic value to a collection of assorted handmade sustainable gifts such as bags, cushion covers, diaries, bangles, textile jewellery, bow ties, bag charms, buntings, hangers, among other such products. For each collection, a range

A   B

Fig. 6 Collaboration of Brand with the Artisans. (A) Traditional printed fabrics (Reprinted under Creative Commons license, Karolia and Buch [19] (B) Contemporised products by Vraj: Bhoomi (*Source* [39, 40])

of blocks is prepared with traditional and contemporary motifs based on a distinctive theme. The printing is done in different colourways depending upon the theme by using their in-house unique block designs for the season on yardage fabrics to make garments or do engineered layouts for sarees, scarves and dupattas. The base fabric and the natural dye ingredients for making the collection are sourced from the Kutch artisans and the entire product except stitching is made in Kutch. The brand has its root in craft and community and is producing more sustainable fashion (Fig. 6).

5   Conclusion

Achieving sustainability is no longer an option but a must. The traditional craft, if reinvented and reinvested can contribute to sustainable fashion. Craft design collaborations enable the craft community to get access to potential markets and new ideas for product development. A push to the crafts built on values can contribute to the sustainable transformation. The craft has to evolve, transform and adapt for meeting the needs of the modern society. The collaborations can play a significant role for revitalising crafts and economic development. The objective of this chapter was to understand the approaches to making fashion more sustainable. Despite complexities in the fashion socio-technical regime, niche players are abandoning the ideals of fast fashion. The case studies depict the possibility of change and can further provide stimulus to practitioners for further research.

References

1. Asheim B, Coenen L, Vang J (2007) Face-to-Face, Buzz, and Knowledge Bases: sociospatial implications for learning, innovation, and innovation policy. Eviron Plann C Gov Policy 25(5):655–670. https://doi.org/10.1068/c0648
2. Barker T, Hall A (2009) GoGlobal: how can contemporary design collaboration and e-commerce models grow the creative industries in developing countries? In: Proceedings of the 3rd IASDR Conference on Design Research. Korean Society of Design Science, Seoul, Korea, pp 2227–2236. https://www.researchgate.net/publication/266183866_GoGlobal_How_can_contemporary_design_collaboration_and_e-commerce_models_grow_the_creative_industries_in_developing_countries. Accessed on 15 Oct 2021
3. Brundtland GH (1987) Our Common Future. Oxford University Press, Oxford. https://www.are.admin.ch/are/en/home/media/publications/sustainable-development/brundtland-report.html. Accessed on 10 Sept 2021
4. Cachon G, Swinney R (2011) The value of fast fashion: quick response, enhanced design and strategic consumer behavior. Manage Sci 57(4):778–795. https://doi.org/10.1287/mnsc.1100.1303
5. Caliope T, Lazaro JC (2018) From fast to slow: transitions to sustainable fashion—a theoretical essay. 9th International Sustainability Transitions Conference. https://documents.manchester.ac.uk/display.aspx?DocID=37303. Accessed 15 Oct 2021
6. Clark WC, Harley AG (2020) Sustainability science: toward a synthesis. Annu Rev Environ Resour 45(1):331–386. https://doi.org/10.1146/annurev-environ-012420-043621
7. Cox E, Bebbington J (2014) Craft and sustainable development: an investigation. University of St Andrews. https://citeseerx.ist.psu.edu/viewdoc/download?doi=10.1.1.933.6413&rep=rep1&type=pdf. Accessed 15 Oct 2021
8. Crabbe A (2012) Three strategies for sustainable design in the developing world. Des Issues 28(2):6–15. https://doi.org/10.1162/DESI_a_00139
9. Dhamija J (2003) From then till now. http://indiatogether.org/craftsnow-economy. Accessed on 10 Sept 2021
10. Dholakia K1, Parmar U (2021, June 1) Design exploration in bead work of Gujarat-India: Experience and outcome of co-creation. Int J Text Fashion Tech (IJTFT) 11(1):61–74. http://www.tjprc.org/publishpapers/2-29-1622290156-7IJTFTJUN20217.pdf. Accessed on 14 Oct 2021
11. Fletcher K (2014) Sustainable fashion and textiles: design journeys. Routledge, Milton Park, Abingdon, Oxon, UK. https://pdfroom.com/books/sustainable-fashion-and-textiles-design-journeys/DkgVebEyd9B. Accessed on 15 Oct 2021
12. Fuchs C, Schreier M, Van Osselaer SM (2015) The handmade effect: What's love got to do with it? J Mark 79(2):98–110. https://doi.org/10.1509/jm.14.0018
13. Geels F, Schot J (2007) Typology of sociotechnical transition pathways. Res Policy 36(3):399–417. https://doi.org/10.1016/j.respol.2007.01.003
14. Granskog A, Lee L, Magnus K, Sawers C (2020) Mckinsey & company Survey: Consumer Sentiment on Sustainability in Fashion. https://www.mckinsey.com/industries/retail/our-insights/survey-consumer-sentiment-on-sustainability-in-fashion. Accessed on 14 Oct 2021
15. Hofverberg H, Kronlid D, Östman L (2017) Crafting sustainability? An explorative study of craft in three countercultures as a learning path for the future. Nord J Sci Tech Stud 5:8. https://doi.org/10.5324/njsts.v5i2.2314. https://www.researchgate.net/publication/322022369_Crafting_sustainability_An_explorative_study_of_craft_in_three_countercultures_as_a_learning_path_for_the_future. Accessed on 15 Oct 2021
16. Jha B (2019) Design Revolutions, Sujni Embroidery and Community of Practice in Bihar International Association of Societies of Design Research Conference. https://iasdr2019.org/uploads/files/ma-f-1098-Jha-B.pdf. Accessed on 15 Oct 2021
17. Job V, Nikitin J, Zhang SX, Carr PB, Walton GM (2017) Social traces of generic humans increase the value of everyday objects. Pers Soc Psychol Bull 43(6):785–792. https://doi.org/10.1177/0146167217697694

18. Jung S, Jin B (2014) A theoretical investigation of slow fashion: sustainable future of the Apparel Industry. Int J Consum Stud 38(5):510–519. https://doi.org/10.1111/ijcs.12127
19. Karolia A, Buch H (2008) Ajarkh, the resist printed fabric of Gujarat. Indian J Tradit Knowl 7(1):93–97. http://nopr.niscair.res.in/bitstream/123456789/578/1/IJTK%207(1)%20(2008)% 2093-97.pdf. Accessed on 14 Oct 2021
20. Markard J, Raven R, Truffer B (2012) Sustainability transitions: an emerging field of research and its prospects. Res Policy 41(6):955–967. https://doi.org/10.1016/j.respol.2012.02.013
21. Murray K (2010) Outsourcing the hand: an analysis of craft-design collaborations across the global divide. Craft + Design Enquiry 2:1–23. https://www.academia.edu/2409566/Out sourcing_the_hand_An_analysis_of_craft-design_collaborations_across_the_global_divide. Accessed on 15 Oct 2021
22. Niinimäki K, Hassi L (2011) Emerging design strategies in sustainable production and consumption of textiles and clothing. J Clean Prod 19:1876–1883. https://doi.org/10.1016/ j.jclepro.2011.04.020
23. Nugraha A (2012) Transforming tradition. Aalto University publication series, Doctoral dissertations. https://aaltodoc.aalto.fi/handle/123456789/20090. Accessed on 15 Sept 2021
24. Padovani C, Whittaker P (2015) Modern alchemy: Collaboration and the value of social capital. Craft Res 6(1):99–111. https://doi.org/10.1386/crre.6.1.99_1
25. Pathak S, Mukherjee S (2020) Entrepreneurial ecosystem and social entrepreneurship: case studies of community-based craft from Kutch, India. J Enterprising Communities: People and Places in the Global Econ 15(3):350–374. https://doi.org/10.1108/jec-06-2020-0112
26. Reubens R (2010) Bamboo canopy: creating new reference-points for the craft of the Kotwalia community in India through sustainability. Craft Res 1(1):11–38. https://doi.org/10.1386/crre. 1.11_1
27. Santagata W (2002) Cultural districts, property rights and sustainable economic growth. Int J Urban Reg Res 26(1):9–23. https://doi.org/10.1111/1468-2427.00360
28. Shiner L (2012) Blurred boundaries? Rethinking the concept of craft and its relation to art and design. Philos Compass 7(4):230–244. https://doi.org/10.1111/j.1747-9991.2012.00479.x
29. Thomas A (2006) Design, poverty, and sustainable development. Design Issues 22(4):54–65. https://doi.org/10.1162/desi.2006.22.4.54
30. Tung F-W (2012) Weaving with rush: Exploring craft-design collaborations in revitalizing a local craft. Int J Des 6(3):71–84
31. United Nations (2016) The sustainable development goals report. https://unstats.un.org/sdgs/ report/2016/. Accessed on 15 Oct 2021
32. Väänänen N (2020) Sustainable craft dismantled and reassembled. Doctoral dissertations, The University of Eastern Finland. https://erepo.uef.fi/bitstream/handle/123456789/22540/ urn_isbn_978-952-61-3319-5.pdf?sequence=1&isAllowed=y. Accessed on 10 Oct 2021
33. Väänänen N, Pöllänen S (2020) Conceptualizing sustainable craft: concept analysis of literature. Des J 23(2):263–285. https://doi.org/10.1080/14606925.2020.1718276
34. Väänänen N, Pöllänen S, Kaipainen M, Vartiainen L (2017) Sustainable craft in practice—from practice to theory. Craft Res 8(2):257–284. https://doi.org/10.1386/crre.8.2.257_1
35. Vartiainen L, Kaipainen M (2012) Textile craft students' perceptions of sustainable crafts. Prob Educ 21st Century 43:131–140. http://www.scientiasocialis.lt/pec/files/pdf/vol43/131-140.Var tiainen_Vol.43.pdf. Accessed on 15 Oct 2021
36. World Commission of Environment and Development (WCED) (1987) Our Common Future. Oxford University Press. https://www.are.admin.ch/are/en/home/media/publications/ sustainable-development/brundtland-report.html. Accessed on 10 Sept 2021
37. Yair K (2010) Activism at work—crafting an alternative business. https://www.historiadeldiss eny.org/congres/pdf/10%20Yair,%20Karen%20ACTIVISM%20AT%20WORK.%20CRAF TING%20AN%20ALTERNATIVE%20BUSINESS.pdf. Accessed on 15 Oct 2021
38. Zhang W, Walker S, Mullagh L (2019) Craft and sustainability: reflections on design interventions in craft sector in China. International Association of Societies of Design Research Conference 2019. Design Revolutions. https://iasdr2019.org/uploads/files/Proceedings/ma-f-1436-Zha-W.pdf. Accessed on 14 Oct 2021

39. https://apparelresources.com/fashion-news/features/vraj-bhoomi-reviving-the-ajrakh-craft-through-treasured-pieces/. Accessed on 15 Oct 2021 (by Anjori Grover Vasesi 22-December-2019)

40. https://vrajbhoomi.in/our-story/

41. https://marigolddiary.com/2016/02/11/swati-kalsis-avant-garde-sujani/

42. http://www.borderandfall.com/people-spaces/swati-kalsi/

Sustainable Fashion: African Visual Artist's Contribution to the New Paradox Discussion—Case of El Anatsui as a Sustainable Thinking Agent

Marc-Arthur Gaulithy, Christine Nantchouang Ngomedje, Prisca B. Pange-Ibinimion, and Ivan Coste-Manière

Abstract For more than two decades, in our current consumerism society, the issue of sustainability has established itself as one of the main paradoxes in the textile and fashion industry. Production methods and consumption habits are questioned, and various initiatives are calling to think sustainability an integral part of every company's business strategy. In order to achieve these goals, various media are solicited to influence the wide variety of actors in the value chain of these industries, to inscribe their actions in this new paradigm. Among these media, contemporary art seems, more and more, to act as emerging potential, in that it constitutes a powerful way to shape societal conceptions. Indeed several contemporary artists and designers behave as change agents, practicing environmental art, which often takes into consideration the broader context of changing societal habits, the origin of the materials used. Such practices could be a real catalyst for a paradigm change. For this, It would be necessary to identify the concepts and mechanisms mobilized by the artists to convey successfully these sustainability values. To do this, in exploratory research, we have studied the case of the Ghanaian world-famous artist, El Anatsui, as well as the impact of his work on different types of actors in the global fashion industry. The results of this case study could be the basis for larger work, which can help accelerate the fashion industry's green transition.

Keywords Sustainability · Fashion · Textile · Sustainable thinking · Change agent · El Anatsui

1 Introduction

For more than two decades, the issue of sustainable development has gradually emerged as one of the most significant challenges of our time. At the center of scientific and organizational debates, this challenge is all the more complex as it not only

M.-A. Gaulithy · C. N. Ngomedje · P. B. Pange-Ibinimion · I. Coste-Manière (✉)
SKEMA Business School, Luxury and Fashion Management, Université Côte d'Azur, 60 Rue Dostoievski, 6902 Sophia Antipolis Cedex, France
e-mail: ivan.costemaniere@skema.edu

S. S. Muthu (ed.), *Sustainable Approaches in Textiles and Fashion*,
Sustainable Textiles: Production, Processing, Manufacturing & Chemistry,
https://doi.org/10.1007/978-981-19-0874-3_10

concerns all industries, but also requires awareness of the need to rethink development, as well as existing economic models. This need for change was recently reaffirmed in the United Nations Agenda 2030, which urges both governments and industries to actively work toward the achievement of sustainable development goals. The creative industries, including the fashion industry, are no exception to this questioning of practices, production methods, and consumption habits [23].

Faced with the complexity of the subject, various media are mobilized to try to influence the various stakeholders (houses, designers, consumers, suppliers of raw materials, etc.) of the fashion industry, when it comes to inscribing their actions and their motivations in the perspective of conservation of the planet. Contemporary Art is one of these means. It is capable, through its ability to conceptualize and rethink everyday life, of being a powerful instrument for raising awareness of the concept of sustainability in the fashion industry [6]. In this approach, one of the names that strikes first, at the global level, as much by his notoriety, as by the originality of his artistic, societal and political approach, as by the importance of the actors of the fashion industry that his work continues to influence, is Nigeria-based Ghanaian artist El Anatsui. Throughout his career, he has been able to mobilize the popular, urban and traditional imagination of West Africa at a crossroads, to educate some of the global players in the fashion industry on the need to resort to a circular economy and new creative techniques.

This article aims to identify, first of all, the concepts and mechanisms that were mobilized in his work to achieve this positioning, but also, to analyze the values and the message conveyed by his work.

Secondly, we will study the impact of his work on the fashion industry in search of a paradigm shift: which designers and houses are influenced? This will allow us to draw some lessons from the El Anatsui case, which could help accelerate the green transition of the fashion industry.

2 Methodology

In this article, we show that the issue of sustainability in the fashion industry must be approached from a multidisciplinary perspective. It is not enough to examine it only from an economic and managerial point of view, but it is also necessary to take into account the contribution that the creative part of the fashion industry can make. To reach this objective, we proceeded, in a qualitative approach, with an exploratory research, by a review of the scientific literature specific to the question of the diffusion of the discourse of sustainability in the luxury and fashion industry, and by the analysis of academic and professional production on the work of El Anatsui. We also proceeded by analyzing internet pages and secondary audio-visual materials found on internet platforms—such as YouTube—about the artist's work and discourse, and reviews.

The analysis of this material has enabled us to identify concepts and a discourse that we have crossed with the results of similar academic work. Of course, one of the original features of this work is that we have embedded it in a narrative based on the contribution of textiles of African origin (mode of production, sociological determinants, perspectives) in the global discussion on the sustainable development of changing fashion industry.

3 Theory

3.1 The Current Issue of a More Sustainable Fashion

In 2009, the Copenhagen Fashion Summit [17], initiated by Eva Kruse, set up the first sustainability agenda for companies in the fashion industry. For the first time, a report addressed to the fashion industry's top executives was published, known as the "Fashion CEO Agenda". The aim of this report, developed in collaboration with groups such as Kering, H&M, and Li & Fung, was to identify seven priority actions to achieve greater sustainability within the industry. These are:

- Traceability of the entire supply chain,
- Intelligent management of water, energy and chemicals,
- Safety and respect for workers,
- The use of sustainable materials,
- A circular mode system,
- Better wage systems,
- The digital transformation of the industry.

As Eva Kruse, CEO of Global Fashion Agenda, explained, "Fashion is one of the world's most important industries, but also one of the most resource- and labor-intensive. The environmental, social, and ethical challenges facing the industry today are not only a threat to the planet, but also a threat to the industry. That's why there is no choice but to make sustainability an integral part of every company's business strategy" (Paraphrased).

The aim of the Fashion CEO Agenda, which has been regularly updated to date, was to encourage fashion industry leaders to lead their design and R&D teams to create more sustainable, with a focus on mixing fibers and producing recyclable apparel. On the other hand, retailers were encouraged to create more used clothing collections. Major fashion companies were encouraged to work with governments to develop better circular systems and to develop innovative technologies to transform textile waste into high-quality fibers.

According to Gazzola et al. despite this increased incentive, the fashion industry remains far from being a circular system, where materials are designed and recycled to generate "*added value*" rather than "*additional waste*" [22].

3.2   Circular Economy and the Fashion Industry

The question of the circularity of the economic model has imposed itself as one of the major strategic issues of the fashion industry. The high number of academic and professional references that speak of it attest to this [22]. It sits alongside supply chain transparency, the use of less polluting materials, and ethical sourcing as the key words of the global strategy to achieve a more sustainable industry.

According to the Ellen McArthur Foundation, "A circular economy is a systematic approach to economic development designed to benefit business, society, and the environment. In contrast to a "take-make-waste" linear model, a circular economy is regenerative and aims to gradually decouple growth from the consumption of finite resources. It is based on three principles: designing out waste and pollution, keeping products and materials in use and regenerating natural systems" [15].

Although this issue may, at first glance, seem to be an additional constraint to the development and maintenance of growth, it is not at all. A report called the "Circular Fashion Report", the result of the collective work of industrialists (Lablaco, Startupbootcamp fashiontech), academics (ESSEC Business School, Wageningen University) and consulting firms (PWC, Anthesis, Rödl & Partner), estimates the market for the circular economy in fashion at 5 trillion USD [29] (Fig. 1).

Fig. 1   Circular Fashion Ecosystem (Circular Fashion Report 2020) Lablaco [29]

In the same vein, many international initiatives bring together the main players in the industry, such as the major brands, the various players in the value chain, and governments. The G20 Circular Fashion Workshop is one of these. Held in June 2021, it also saw the participation of delegations of young entrepreneurs from the South, who, among other things, called for a fashion industry whose future business models take into account human rights over profits, environmental conservation over production, and circularity over linearity. According to them, the time has come to drive a strong commitment between civil society, different levels of governments, and the private sector, to reduce the industry's carbon footprint through inclusive policies. They also called for a revaluation of the cost of labor, whose weakness predominantly impacts women in the South [21].

In this circular economy called for by all, African countries are not left out. Although they represent only a tiny part of the world's consumption of luxury goods, they are nonetheless important players in other links of the value chain. Indeed, they now concentrate a significant part of the producers of raw materials such as certified sustainable cotton Made in Africa (CmiA), whose production was estimated at 578.562 mt of fiber in 2017/2018, i.e., 2.17% of global cotton production [40]. The African continent is also one of the main territories for relocation of manufacturing due to the low cost of labor [35].

3.3 West African Traditional Textile (Kente)

Through countries like Benin, Ivory Coast, Burkina Faso, and Mali, West Africa is considered one of the largest cotton producing regions in Africa. Both white and brown cotton are produced. This production is estimated at 5% of world production, or 2.3 million metric tons [38]. This cotton is exported to other regions, notably to Asia, which remains the leading importer of cotton made in Africa. However, local consumption has been established since ancient times, and consisted of the manufacture of traditional handmade textiles, without recourse to chemical dyes. This know-how, which was passed on from generation to generation, and whose custodians were generally women, organized in communities, generally consisted of woven materials, totally handmade and dyed with natural dyes [9]. Each tribe has developed its own textiles, easily recognizable among a thousand. These textiles of African origin, by their texture, their patterns and their often highly colored characters, have inspired several world-famous fashion designers, such as Jean-Paul Gaulthier, John Galliano, Alexander McQueen, and have also been worn by celebrities such as Beyoncé, Kanye West or Rihanna [4].

One of the most symbolic of these textiles of African origin is the Kente. Originally from Ghana, the Kente is also produced in Ivory Coast. According to Boateng's work, there are two main types of Kente [9]. The Asante Kente and the Ewe Kente, Asante and Ewe being two tribes living in present-day Ghana and Côte d'Ivoire.

According to this work, Asante Kente is made up of single woven strips with alternating double woven panels so that when the strips are sewn together, the effect is

Fig. 2 Asante Kente, The King has Boarded the Ship (1985, Rayon). Photo by Courtnay Micots, 2017

similar to a checkerboard. It is distinguished by its bright colors and abstract patterns woven into the strips, unlike Ewe Kente in which the colors are more subdued and the patterns more realistic (see Fig. 1). The patterns used in the weaving of Asante Kente cloth have specific names; however, the cloth is usually named after the colors and the background pattern, which is often striped. Kente may also be named after historical figures and events as well as Asante values. For example, the design "kyeretwie", or leopard catcher, symbolizes courage, while "aberewa ben", or "wise old woman", indicates the respect given to older women in Asante society. While the designs of Kente cloth are generally abstract, weavers have expanded their aesthetic frameworks by weaving elements such as words, numbers, and Adinkra symbols into cloth. These are usually strips intended for use as stoles or as decoration (Fig. 2).

We see that Kente is not only a textile that is valued for its high aesthetic quality, but also because it is a medium to convey values and messages. It is also important to note the close link that the artisanal production of certain Kente designs had with the Asante royalty. Indeed, some craftsmen were reserved for the production of Kente only for the use of the Asantehene [King of the Asante] [9]. This also makes it a conveyor of prestige values. Although today, this exclusivity tends to no longer exist, these particular designs of Kente, which legend has it that a woman created, remain some of the most expensive textiles in Africa, and highly symbolic of wealth, nobility, high social status, and cultural sophistication, which are given as gifts at important ceremonies such as weddings. These fabrics are then kept as an inheritance that is passed on from mother to daughter or from father to son [9, 12].

3.4   Cultural Appropriation

The symbolism conveyed by textiles of African origin is rich in diversity and meaning. As we have seen above, it conveys multiple cultural values, but not only. They carry political messages, notably of identity, colonial and postcolonial resistance, but also of ethnic affirmation of a proud Africanity in a globalized world. The first President of Ghana, Kwame Nkruma, during his visit to the USA in 1958, appeared alongside President Eisenhower wearing Kente. This made headlines in newspapers and magazines all over the world. Since then, in the USA, the Kente has become an important symbol of African identification for Afro-descendant Americans. It is also used for graduation or other honorary ceremonies [9].

It is not uncommon today to see some stars of the international showbiz world or black American or European political personalities wearing clothes made of African textiles. These creations are often the work of designers of African origin, but also of major fashion houses. It has even become a testimony of return to the sources, an act of militancy. However, the presence of Africa in the global fashion industry scene is not new. It goes back several decades. In an article entitled *"The idea of Africa in European High Fashion: Global Dialogues"*, Kristyne Loughran takes a retrospective look at this permanent search for exoticism in European high fashion, and more precisely, at the presence of African textiles, patterns, and shapes in designers' creations for centuries. Focusing on the year 1997, she shows how the world of Haute Couture has often been inspired by the symbolism conveyed by textiles of African origin [30].

In a globalized world, where cultural boundaries are breaking down with technological innovations and revolutions, and where aesthetic standards are becoming more democratic, it is common practice for designers to transfer trends belonging to one culture into attractive designs for use by other cultures. From the point of view of fashion historians, this recourse to "African exoticism" in the sphere of Haute Couture, by designers such as Coco Chanel, Elsa Shiaparelli, Paul Poiret, and Sonia Delaunay, is done via several mechanisms such as borrowing, appropriation, reinterpretation… [25, 30].

According to Scafidi, cultural appropriation is defined as «*taking from a culture—not our own—intellectual property, cultural expression or artifacts, history or know-how*» [36].

All aspects of culture can be appropriated; be it religious practice, language, folklore, cuisine, and any other type of cultural identity constituent. However, this does not fail to raise the question of how? Some West African textiles such as Adinkra or Bogolanfini were used for specific rituals and ceremonial contexts such as funerals or initiation rites. In the process of reinterpretation or borrowing by which some designers have brought them to the T's of the haute couture and luxury industry, some conscientious objectors have questioned the possibility of cultural dispersion and corruption of the message and values conveyed. While it is understood that most of the time, haute couture designs convey specific messages, the issue of the diversion of the cultural and cultic value of the symbolism of these textiles is not trivial. According to Buchloh, any attempt to reappropriate historical patterns should be motivated by a desire to establish continuity in the tradition, with some acceptable modifications, for the sole purpose of enabling universal application. Such an approach would hardly be questionable. To do otherwise could lead to challenges and the reintroduction of accusations of spoliation and other types of conflict that often pit powerful, hegemonic majorities against powerless minorities [11]. This is what Xanthe Brooks refers to in "*Why Borrowing Should Never Tip Into Exploitation*" [10]. Such a discourse tends to portray any Western designer or Western fashion house, which borrows or appropriates traditional African textiles, as part of a dominant group that exploits a culture different from its own.

The use of cultural symbols should be done with a good comprehension of the values and tradition of these symbols to ensure that they are used appropriately and for the benefit of those to whom they belong. It is not enough for designers to think that by simply using these symbols they are honoring their owners [1].

Another type of issue this raises is that of intellectual property, which is difficult to delineate and address, especially since these designs, textiles, artifacts and their history belong to the intangible heritage of entire communities. According to Arewa, understanding the context in which borrowing, or cultural appropriation, occurs is important in order to avoid any appropriation that would be tantamount to exploitation. He believes that it should be placed within a legal, institutional and commercial framework [2]. Boateng does not argue otherwise in his timely book "*The Copyright thing doesn't work here*" [9].

3.5   Contemporary Artists and Designers as Change Agents

Since the Copenhagen Fashion Summit in 2009, the issue of sustainability has become a central issue in the industry. Seven priority actions have been identified in order to achieve sustainable production and consumption of products. Designers and other creative craftsmen are central to the luxury industry, in that they have the heavy and exciting responsibility of creating innovative and unique pieces, which at the same time reflect the uniqueness and prestige of the brands, breathe contemporary approaches, and embrace the most important societal issues such as the need for a circular economy. They are, therefore, by virtue of their position and the artistic nature of their approach, a force of proposition and a vital contributor to raising awareness of the paradigm shift [13, 20].

Indeed, culture as a multidisciplinary sphere in which humans think, act freely, and shape new worldviews is a powerful vehicle for societal transformation and innovation. According to Trompenaars and Hampden, the concept of culture manifests itself as a demonstration of a people's identity. It is a way of showing how that people influence numerous fields of art and design, including music, painting, sculpture, textiles, and fashion [41]. It can therefore be argued that culture, and the creative people who are artists, are a powerful agent of societal change. Whether they initiate them, or whether they echo them much more widely, their contribution is undeniable.

With this approach in mind, in 2013 and 2014 there was widespread support for a campaign to include culture as part of the Sustainable Development Goals. This campaign was supported by the International Federation of Arts Councils and Culture Agencies (IFACCA) and other key international networks, all that was in collaboration with UNESCO [27]. The campaign highlighted the advantages of paying more attention to the fundamental role of culture as a development outcomes accelerator. This was reiterated at the 2019 meeting on the sustainable development agenda.

According to Moore and Tickell, most artworks are designed with a target audience in mind, even when their ultimate impact is much greater [31]. Thus, so-called environmental art, which often takes into consideration the broader context of changing societal habits, the origin of the materials used, and the ecological impact of how an artwork was constructed and disseminated, can have effects that extend over time on lives and organizational systems.

More specifically, there is an academic on art and design that argues that artists can act as agents of change in innovation processes [32, 34]. Although artists and designers can play such a role in innovation processes in many different fields, there is some that focuses on innovations in the context of sustainability.

According to Barnard et al., it is important to rethink the traditional division between the disciplines of art and design in order to better understand the role of artists and designers as agents of change. Fine art and design can no longer be considered as completely separate domains. Designers often also work as artists, and artists frequently also design as part of their practice [5].

Indeed, in the production of media art, artists usually take on both design and technical responsibilities, and regularly switch between the roles of artist, designer, and technical developer throughout the process [33].

4 El Anatsui: A Work, a Commitment to Paradigm Shift

The contemporary art of African origin has known for some years a considerable rise. It is also exhibited in the most important galleries and festivals such as the Venice Biennale. One of the faces that most embodies this international influence is that of the Ghanaian visual artist El Anatsui. A product of the post-independence art movements of the 1960s and 1970s, the artist has managed to rise to the top of the contemporary art market as one of the most significant living African artists [26]. The Ghanaian sculptor living in Nigeria continues to influence the entire art world in all disciplines with his powerful and unique artistic expression. From sculpture-installation to fashion design, through photography, several contemporary artists have appropriated his work. His work is deeply rooted in the Ewe tradition (a tribe from Ghana), soaked in the hustle and bustle of the African metropolis, and finally resolutely turned toward a societal paradigm shift.

A product of the post-independence West African art movements, this visual artist is universally acclaimed for his ability to manipulate rigid material into sculptural installations reminiscent of one of Africa's most famous textiles: Kente. In doing so, his work is often associated with that of textile designers.

El Anatsui's wall sculptures are generally monumental, glittering, cascades of pixelated metal that at first sight seem to be very sophisticated, but are made from recycled twist-off thin and small metal of all kinds and colors.

They are most of the time made with recuperated waste such as liquor bottle caps, evaporated milk tops, rusty metal rasps, and old printing plates. As you may understand, he uses materials readily available in his immediate environment. In his hands, these materials are mastered as mediums to manufacture complex artworks that seem like typical West African colorful fabrics. In El Anatsui's hands, these various materials clearly reveal their potential as artistic media. After being cutted, rolled, twisted, crushed, and then stitched together with copper wire [42] (Fig. 3).

Fig. 3 El Anatsui, Red Block, 2010. Courtesy: Broad Art Foundation, Brooklyn Museum

It is reminiscent of the abilities that were attributed to alchemists in medieval Europe: the ability to transform vile material into gold.

As noted in the previous chapter, the large Kente cloths from which Anatsui's works are inspired were the prerogative of the powerful in the cultural tradition of Ghanaian weavers. They also convey values and are used to convey certain messages. By mimicking the Kente, the creations of El Anatsui are far from being fortuitous. In doing so, because he confers to the hard and arid metal the suppleness of the fabric, the community of Nsukka where he lives and works, lends to El Anatsui the occult powers that are recognized to the blacksmiths for their mastery of fire and their capacity to metamorphose the elements. Hence his nickname of "alchemist blacksmith", given by some critics. He knows how to transform iron from Nsukka's dumps into gold. El Anatsui is a transformer who spreads the message of change.

In a publication entitled "Fashion as performance: Influencing future trends and building new audiences", author and cultural producer Nicole Shivers brings together fashion designers such as Oumou Sy, Lamine Kouyaté (Xuly Bët), and Alphadi, with performance art, particularly those made by artists of African origin such as El Anatsui in his GAWU installation exhibition. The common practice of *"Making Something out of Nothing"* emerges [37]. This approach, both metaphorical and transformative, constitutes a sort of red thread in the artist's work.

On this subject, El Anatsui will say:

> When I create a work, I see it as a metaphor reflecting an alternative response; examining the possibilities and pushing the boundaries of art. My work can represent connections in the evolving narrative of memory and identity. The link between Africa, Europe and America is very much behind my work with bottle caps. I have experimented quite a bit with some materials. I also work with material that has witnessed and encountered a lot of human touch and use… and this kind of material and work has more charge than the material/work I have done with machines. Art is born out of each particular situation, and I think artists benefit from working with whatever their environment offers. [28]

As we can see, the artist's work can be interpreted with a double meaning. Asenso et al. [3] argue that El Anatsui's work is very militant and philosophical. By using discarded materials such as using bottle caps of old and new liquors, which for some, carry with them the postcolonial history of African countries, and for others, are a symbol of urban cities consumerism, El Anatsui constructs a discourse that is both metaphorical and militant. He shows his interest in recycling, transformation, and an intrinsic desire to connect with each human being while telling his part of the truth about the relationship between them, their waste and their environment. This has earned him, in the opinion of many, the label of the environmental artist.

El Anatsui's work challenges, in a different but more beautiful form, the previously discussed issues of appropriation and borrowing. He himself comes from the communities from which the traditions and artifacts he mobilizes originate, but he uses them to address a discourse that is sometimes more political and global.

El Anatsui's work is unique in that it is deeply rooted in the community. He doesn't just pretend to speak for the Ewe or Nsukka communities. It involves them in its realization. Indeed, El Anatsui's huge workshop in Nsukka, Nigeria, is teeming with people, skilled artisans, and less skilled hands. The work is truly participatory. From the collection of raw materials in the garbage dumps to their transformation into huge fabrics of iron, copper, and others. The creation of his monumental pieces mobilizes several dozen people for dozens of days [26]. This allows the care of hundreds of people behind these dozens of families. El Anatsui's creative activity is clearly part of a circular economy approach. Perhaps this is, beyond the great aesthetic value of his designs, and the political discourse he relays, also what seduces and impacts the creative community in the world of fashion and high fashion?

5 Impact in the Global Fashion Industry

Rarely has the name of a visual artist and sculptor been mentioned so often in the fashion and haute couture world. In addition to the fact that the general aesthetic of his installations resembles Kente or Adinkra (two textiles originating from Ghana), the play of volumes with the material, the shapes he gives it, this sensation that these "metal fabrics" seem so malleable, have inspired many designers and haute couture collections.

It is well known that the line between performance art of this kind and the world of fashion is so fine that one would only realize it once it has been crossed. If pieces such as "The Man's cloth" had opened the way, the performance exhibition "Gravity and Grace" completed the conviction that there is almost no difference between the quintessence of his work and haute couture creations.

5.1 TSE

Founded in 1989 by Augustine Tse and headquartered in New York, TSE is a cashmere brand, supporting the idea that this luxury textile, until then, had been largely underutilized, being perceived as being reserved for the elite. The company's goal was to integrate this incredible material into all aspects of one's life and make it more accessible to a wider audience. It was this modern adaptation of the traditional product that changed the luxury cashmere business forever. The company remains true to its original philosophy of continuing to be a laboratory of ideas and innovation.

Firmly committed to a circular economy and make trade, TSE controls the entire value chain of raw material production. In collaboration with the Xiang Jiang provincial government in China, TSE has full control, from securing the purest raw materials to spinning operations to the production of all finished products. From the beginning, TSE has been committed to consistently providing quality products [18].

On the Spring 2018 collection page, the narrative to introduce it notes that TSE's design team, in its ambition to reinterpret the art of knitting, produced "raw" textured coats inspired by the conceptual work of El Anatsui. Specific mention is made of one of his solo exhibitions in New York, at the Brooklyn Museum, or Instagram views of his installations:

El Anatsui connects thousands of caps with copper wire to form "sheets" that take over an entire wall, and depending on how they're strung up-he lets the gallery or museum hang them as they please-they look like dripping metal. If you're fashion inclined, they look like fabric. How could a knitwear designer not be inspired?

Where other labels might have seen an opportunity to re-create the bottle tops as a print or big sequin, TSE's designers were interested in Anatsui's process as well as the "malleable" nature of his work. Dense, stretchy knits featured delicate, hand-linked squares, for instance, and hand-crocheted circles of yarn formed airy layering pieces. The palette was drawn from Anatsui's work, too, like rich persimmon, copper (which was almost shiny on mercerized cotton), and petal pink, best seen on the knit culottes and tube tops. Then there were the chunky cardigans and poplin skirts that "collapsed" around the body, as well as curve-gliding satin slip dresses. In the past, TSE collections often featured just a few non-knit items, but Spring is a study in how the TSE customer really wears her knits. She might style the slip dress by itself or under the crochet knit, and there are many other ways to wear it, too. Like an Anatsui piece, they left room for interpretation. [43]

5.2 Sustainable Thinking

"Sustainable Thinking" is a project held in 2019, thanks to the Salvatore Ferragamo Foundation, under the auspices of the eponymous House.

It proposed exhibitions and collateral initiatives involving the Salvatore Ferragamo Museum in Florence and other public institutions in the city. The aim of this project was to suggest reflections on a theme so important for the future and a paradigm shift, at least in the world of fashion, art and architecture.

Several disciplines took part in this high-profile event. Among the participants were renowned architects, fashion designers such as Paul Andrew, Lucia Chain, Maria Sole Ferragamo, Eilen Fisher…and fiber innovators such as ECOTEC, EVO, Re.VerSo … [19].

The Ghanaian artist El Anatsui was also part of the event. He brought his experience and vision of sustainable fashion. The work that was exhibited at the Museo Salvatore Ferragamo is "Energy Spill (2010)". In addition to its dominant colors of gold and red, on closer inspection the visitor can see fragments of cans and other debris. The logos and brands of spirits on the bottle caps can be read: these multiple elements are arranged like checkerboards in a precious contemporary mosaic, linking Africa to the West and, at the same time, past, present and future.

The folds and draperies that result from his installations, which border on the tapestry, transform spontaneously during their creation and change according to the location, giving rise to sophisticated and seductive installations that are reminiscent of the work of designers.

Finally, it is interesting to recall that, according to the promoters of this project, for Salvatore Ferragamo "investing in sustainable development means believing that the use of innovative materials, the link with the local community and the attention to the environment and to people are the key to success. This is perhaps the best way to follow the values passed on by the Founder who, since the 1920s, has been experimenting with natural and unusual materials developing a philosophy aimed at the physical and psychological well-being of his customers".

In addition to these two examples that clearly demonstrate the status of an agent for a paradigm shift for a sustainable fashion industry that can be attributed to El Anatsui, it is also worth mentioning the impact he continues to have on young and promising industry players. Among them, Zoran Bodric [14] and Noél Puéllo. The latter, in response to a question about his sources of inspiration, answers:

> To be 100% honest, I don't think I look at fashion much for inspiration for my design… I went to school to study fibers and materials, not fashion, so my inspiration comes mainly from artists like David Hammons, El Anatsui, Sonya Clark, and Doris Salcedo, artists who are interested in manipulating abundance, recycled materials, and people who want to question the way you see an object. I want to challenge the way we see fabric on bodies, going beyond the realm of just design [39].

> This influence also extends to young fashion editorial writers like Antonio Bertolino who sees a parallel between El Anatsui's work and Dolce & Gabbana's unforgettable Fall 2013 collection. According to him, El Anatsui's pieces from his cultural heritage (Kente) are similar in composition to mosaics in the way the artist creates a large-scale work from tiny pieces. It is this same mosaic aesthetic found inside Byzantine cathedrals that inspired the prestigious Dolce & Gabbana collection. Therefore, for him, it is obvious that both El Anatsui and Dolce and Gabbana use their heritage as a source of creative and compositional inspiration in their work [8].

6 Conclusion

At the root of this article was the question of how the creative link in the fashion value chain could contribute more effectively to the achievement of sustainability goals in the industry, but more specifically, how designers of African origin participate in the discussion as agents of change. Through the example of El Anatsui, we can argue that the African imagination is full of resources that can feed the pipelines of sustainable thinking in the fashion industry on a global scale. If only by thinking a little more in terms of "making something from nothing", and by integrating a resolutely circular approach to the supply chain. Another of the lessons that the El Anatsui case confirms is that the use of education through art is a powerful instrument for paradigm change, in that it is collaborative in nature, and creativity brings into existence processes that did not exist before. Indeed, artistic creation is an approach that easily addresses complexity, is familiar with metaphors and allows us to break out of what Hamel and Prahalad call sectoral orthodoxy [24]. Indeed, even under optimistic assumptions, existing industry solutions and business models will not produce the impact needed to transform the industry in a profound and systemic way.

To achieve this, the industry must constantly innovate and involve as many stakeholders as possible. This is particularly important for the most difficult stages of the value chain: raw materials and end-of-use.

Bibliography

1. Acquaye R (2018) Exploring Indigenous West African Fabric Design in the Context of Contemporary Global Commercial Production. Thesis for the degree of Doctor of Philosophy, University of Southampton, Faculty of Business, Law and Art, Winchester.
2. Arewa O (2016) Cultural appropriation: when borrowing becomes exploitation. Retrieved 15 Oct 2021, from http://theconversation.com/cultural-appropriation-when-borrowing-becomes-exploitation-57411
3. Asenso K, Issah S, Som EK (2020) The use of visual arts in environmental conservation in Ghana: The case of Adams Saeed. J African Art Educ 1(1):66–83
4. Aziz M, Salloum C, Alexandre-Leclair L (2019) The fashion industry in Africa: A global vision of the sector. In: Moreno-Gavara C, Jiménez-Zarco A (eds) Sustainable fashion empowering African women entrepreneurs in the fashion industry. Palgrave Studies of Entrepreneurship in Africa.
5. Barnard B, Van Dartel M, Beekman N, Linderman KP, Nigten A (2015) Artists and designers as agents of change. Participatory Innovation Conference (PIN-C) 2015: reframing design. The Hague University of Applied Sciences, The Hague, Netherlands.
6. Beekman N (2015) The artist as agent of change. In: Nigten A (ed) A sense of Green
7. Benjamin W (1968) Illumination. Boston, USA.
8. Bertolino A (2017, April 3) Art-à-Porter: Dolce and Gabbana, and the celebration of culture in fashion. Retrieved 15 Oct 2021, from The Tufts Daily. The Print Edition: https://tuftsdaily.com/arts/2017/04/03/art-a-porter-dolce-and-gabbana-and-the-celebration-of-culture-in-fashion/
9. Boateng B (2011) The copyright thing doesn't work here. Adinkra and Kente Cloth and Intellectual Property in Ghana. University of Minnesota Press, Minneapolis
10. Brooks X (2016) Cultural Appropriation. Mix Magazine 45, p 17
11. Buchloh B (2009) Parody and appropriation in Francis Picabia, Pop and Sigmar Polke. In: Appropriation. Whitechapel Gallery Ltd, London, p 178
12. Contemporary African Art (2016) Kente Cloth. Retrieved 15 Oct 2021, from https://www.contemporary-african-art.com/kente-cloth.html
13. Debeli D, Jiu Z (2013) Analyzing the cultural background of textile designers' on their innovative thinking. In: International Conference on Education Technology and Management Science (ICETMS 2013), pp 1239–1242
14. Designer Focus (2020, June 18) Designer focus: Zoran Dorbic. Retrieved 15 Oct 2021, from notjustalabel.com: https://www.notjustalabel.com/editorial/designer-focus-zoran-dobric
15. Ellen Macarthur Foundation (2017) Ellen Macarthur Foundation/Archive. Retrieved 14 Oct 2021, from https://archive.ellenmacarthurfoundation.org/explore/the-circular-economy-in-detail#:~:text=A%20circular
16. Enwezor O, Okeke-Agulu C (2009) Contemporary African art since 1980. Damiani, Bologna, Italy
17. Fashion Network (2018, March 27) Global Fashion Agenda: Seven priority actions for sustainable fashion. Retrieved 14 Oct 2021, from https://fr.fashionnetwork.com/news/global-fashion-agenda-les-sept-actions-prioritaires-pour-une-mode-durable,962437.html
18. FMD (1998) TSE. Retrieved 15 Oct 20121, from https://www.fashionmodeldirectory.com/brands/tse/
19. Fondazione Ferragamo (2019, April) Sustainable thinking. Retrieved 14 Oct 2021, from https://www.ferragamo.com/resource/blob/423268/f2178b01483702703a8ec90703b02620/2019-sustainable-thinking-en-data.pdf
20. Frascara J (2002) Design and the social sciences: making connections. Taylor&Francis, London and New York
21. G20, European Commission (2021, June 7) G20 Circular Fashion Workshop. Retrieved 14 Oct 2021, from https://ec.europa.eu/environment/international_issues/pdf/G20%20Circular%20Fashion%20Workshop%207%20June%202021-Full%20Report.pdf

22. Gazzola P, Pavione E, Pezzeti R, Grechi D (2020) Trends in the fashion industry: the perception of sustainability and circular economy: a gender/generation quantitative approach. Sustainability [online] 12(7):2809.

23. Global Fashion Agenda, BCG (2019) Pulse of the Fashion Industry. Retrieved 14 Oct 2021, from http://media-publications.bcg.com/france/Pulse-of-the-Fashion-Industry2019.pdf

24. Hamel G, Prahalad CK (1994) Competing for the future. Harv Bus Rev 72(4):122–128

25. Hannel S (2006) "Africana" textiles: imitation, adaptation, and transformation during the Jazz Age. Textile 4(1):68–103

26. Harney E (2017) El Anatsui's abstractions: transformations, analogies and the new global. In Iskin R (ed) Re-envisioning the contemporary art canon

27. IFACCA (2013, September) IFACCA Report N15. Retrieved 14 Oct 2021, from https://ifacca.org/media/filer_public/86/03/86039fa5-797c-4345-a0ec-1cc3539e6d56/report15fr.pdf

28. James LL (2008) Convergence: history, materials, and the human hand-an interview with El Anatsui. Art Journal 67(2):36–53

29. Lablaco (2020) Year Zero. Circular Fashion Report 2020. Retrieved 14 Oct 2021, from https://docsend.com/view/63avn4jc3ztb952w

30. Loughran K (2009) The idea of Africa in European high fashion: global dialogues. Fash Theory 13(12):243–272

31. Moore S, Tickell A (2014) The arts and environmental sustainability: an international overview. D'Art Topics in Arts Policy, (No. 34b), Julie's Bicycle and the International Federation of Arts Councils and Culture Agencies. Retrieved Oct 2021, from http://www.ifacca.org/topic/ecological-sustainability/

32. Nigten A (2013) Real Projects for Real People, vol. 3, Nigten A (ed). The Patching Zone

33. Nigten A (2014) The Design Process of an Urban Experience. HCI (Vol. 21). Spinger Verlag

34. Nigten A, Van Dartel M (2013) Explorations of ecological autarky in art, design and science, proceedings ISEA International. University of Sydney, Sydney

35. Nordås H (2004) The global textile and clothing industry post the agreement on the textiles and clothing. WTO Discussion Paper No. 5

36. Scafidi S (2005) Who owns culture? Appropriation and authenticity in American law. Rutgers University Press, Brunswick, NJ

37. Shivers ND (2011) Fashion as performance: influencing future trends and building new audiences. In: Fashion Forward. Brill, pp 405–417

38. Sika Finance (2020, July 9) Cotton: Benin 1st cotton producer in Africa. Retrieved 14 Oct 2021, from https://www.sikafinance.com/marches/coton-le-benin-1er-producteur-d-afrique-avec-une-production-de-pres-de-715-000-tonnes_23164

39. Smith A (2019) Thrifting a way to re-imagined fashion, a conversation with emerging designer Noél Puéllo. Retrieved 15 Oct 2021, from https://www.theartblog.org/2019/10/thrifting-a-way-to-re-imagined-fashion-a-conversation-with-emerging-designer-noel-puello/

40. Textile Exchange (2020) 2025 Sustainable Cotton Challenge. Second Annual Report 2020. TextileExchange.org

41. Trompenaars F, Hampden-Turner C (1997) Riding the waves of culture. Nicholas Brealey Publishing, London

42. Vogel SM (2021) El Anatsui: Art and Life. Revised and Expanded. Prestel

43. Vogue Runway (2018) TSE Spring 2018 Ready-t--wear. Retrieved 16 Oct 2021, from https://www.vogue.com/fashion-shows/spring-2018-ready-to-wear/tse

An Old Approach Sees Major Revival: Local Textiles Go Global

Carmelo Balagtas, Allison Elisabeth Stanley, Francois Le Troquer, and Ivan Coste-Manière

Abstract As sustainability is put at the front of every communication strategy of almost any brand who would like to keep themselves relevant for the next decades or so, the trend to seek means on how much is achieved with respect to its three main pillars, environment, economy, and ethics, continues. Now bounded by the Sustainability Development Goals for this decade, brands continue to work alongside with scientist to expedite ways to reinvent traditional operation models. Focused on the Fashion industry, who's legacy of being a major contributor to waste, has seen a bright light to turning tables around. More specifically how it produces its product from clothes to accessories, by means of re-investing both time and resource on natural textiles. This abstract wishes to highlight opportunities found in interesting nations from Nigeria to the Philippines where respective countries' rich natural diversity becomes an enabler to this feat. Take, for example, Nigeria. Who would have realized that the fashion industry would turn heads over to this nation, as it simply prides itself to a textile that is slowly finding its glory back, the 'Tapa'? The material is a fibrous textile common in West Africa. As it is a traditional material, culture and tradition are attached to it. It is believed to be used to honor the Gods and as it is a delicate fabric, its connection to social classes is as well evident to date. Sindiso Khumalo—Sustainable textile designer and LVMH Prize Finalist 2020, is a clear feat that energy exemplified in the region toward such endeavor is not only making waves but are also clearly recognized by fashion houses, and that altogether speaks volumes on how such is a scalable initiative. Meantime in the east, the Philippines as one of the most bio-diverse countries in the world, is abundant in plants and the

C. Balagtas (✉) · A. E. Stanley · F. Le Troquer · I. Coste-Manière
SKEMA Business School, Lille, France
e-mail: carmelo.balagtas@skema.edu

A. E. Stanley
e-mail: allisonelisabeth.stanley@skema.edu

F. Le Troquer
e-mail: francois.letroquer-ext@skema.edu

I. Coste-Manière
e-mail: ivan.costemaniere@skema.edu

© The Author(s), under exclusive license to Springer Nature Singapore Pte Ltd. 2022
S. S. Muthu (ed.), *Sustainable Approaches in Textiles and Fashion*,
Sustainable Textiles: Production, Processing, Manufacturing & Chemistry,
https://doi.org/10.1007/978-981-19-0874-3_11

likes, which turn out to be an effective textile alternative. 'Piña' and 'Abaca' to name a few were once only used for traditional costumes and have found its perfect staging in the Fashion scene. Taken from the Pineapple fibers, 'Piña' is known to be stiff, translucent, lustrous, and expensive. It was favored among the affluent back in the day. On the other hand, 'Abaca', also referred to as, Manila Hemp is derived from the fibers of a banana species native to the Philippines. While the fiber is commonly used for paper and furniture, it is also popularly used in the fashion industry to create dresses, tops, and even accessories. It seems only fitting that the world follows the lead of these two nations or their contemporaries, as it is fitting for obvious reasons to proceed with fashion using the natural alternatives that are compact with sustainability rational. However, storytelling as well as a good angle to consider as it is an equally important component for Luxury and Fashion. On the luxury part these materials produced and developed by artisansremain strong contenders for exquisite quality while on the other side where story telling is vital, the historical past of these cited materials is incredibly astonishing. These traditional textiles align with the Philosophies of sustainable approach to textile best fit for the Fashion Industry.

Keywords Innovation · Sustainability · Textile · Natural textile · Textile alternative · Fashion industry · Luxury industry

1 Introduction

The global textile industry impacts the entirety of the human population. From textile production, refinement, and retail garment sales to textile distribution, consumption, and use, the networks are never ending. In order to examine the current industry, and compare local textiles to global textiles, we must identify the current state of the industry.[1]

Textiles are fundamental to our society and the fashion industry at large. The textile industry is valued at an estimated USD 3 trillion which involves from the end to end process of producing, refining and selling of both synthetic and natural fibers which in turn are utilized in thousands across industries. And to run this operation, millions of labor are necessary. In fact, the textile industry is one of the most demanding in temps of employing people. And given how fundamentally the industry is important to economies, particularly Europe's manufacturing and Asia's rising demand, what is often overlooked in the industry are the hidden costs to the environment. Production and consumption naturally have a significant impact on the environment. Social impact is inevitable as cited above which involves people, in addition is the resources used like water, land, and others that will directly or indirectly produce greenhouse gasses and other pollutants.

The global fashion industry with regard to textiles is changing due to devaluation of design, economic pressure and consumer integrity. In other words, the ethical and unethical ways in which textiles are extracted and used in the fashion industry is

---

[1] Bick et al. [1].

becoming more and more prevalent. Traditional business models are going through some paradigm shifts, where high value is sought from product, service and now with the ethics behind all the work. Mainly driven by the new generation of customers, Millennials, and Gen-Z, these consumers expect nothing less from co-creation, innovation, and personalization of their purchases. This being said requires a level of integrity and transparency from business to the supply chain, to effectively build a relationship between the brand and its patrons. During this chapter, we will provide an overview of geographical textile production, examine the fast fashion industry, outline specific textiles that are deemed unsustainable and sustainable, and then share specific examples and countries that have shifted from global to local textile production and consumption. The purpose of this chapter is to provide a past, present, and future perspective on the textile industry, and discuss the shift to local textiles that is not only inevitable, but imperative.

1.1 Fast Fashion Under the Limelight

Fast Fashion as its words would suggest is a model popularized by 'Zara', where the production of latest trends shown at fashion shows are produced in big quantities and are quickly distributed from catwalk to display racks. Such a model effectively democratizes the ever-changing trend on clothes as consumed by those who have an appetite to be in trend without the need of spending too much. This business model is supplemented by the celebrity and social media culture that exposes them on mainstream TV and through more modern platforms, online, where the appearance and presentation of brands are seen on a variety of personalities.

Apart from Zara, fast fashion companies use cheap textiles commonly sourced from developing countries to produce garments in high street stores that copy the latest fashion trends.

This consistent display of fashion culture stimulates the need to always look new if not different. So much that outfit repeating could be frowned upon. Coupled with the new generation of consumer, and their lifestyle lived online, this could very much be true for the most part. This, however, in turn produces a toxic system of over production and consumption if not mitigated with some form of Artificial Intelligence. Which is the case, and by which made the entire Fashion industry, rank one of the most polluters in the world. In the eyes of the consumer and retailer, fast fashion has been seen as positive due to clothing becoming cheaper and trend cycles speeding up and shopping becoming a hobby. Suddenly, a new fashion era was born where middle-class consumers have access to the latest style and trends at their fingertips for affordable prices. The modern retail fast fashion stores sell cool, trendy clothing that can be purchased for extremely low prices, worn a handful of time, then disposed with the same week or less. In theory, the average person could dress very similar to personalities they follow, wearing a very similar ensemble that mimics designer brands. This global concept of efficient fashion turnover was too good to be true and did not come without consequences. Let us dive into the fashion key players and business models to further understand.

1.2 Fast Fashion Main Stays

Famous fashion retailers include Zara and H&M whose original past resembles quite similar where both brands started modest if not small while building its network in Europe. H&M is the oldest of the fast fashion giants, having opened in 1947 in Sweden that later expanded its distribution networks in London in 1976. Followed by Zara, hailing from Spain where it found its placement in New York at the beginning of the '90s. This became the beginning of the new term, 'Fast Fashion'. The word was coined by the New York Times magazine to describe what the brand was able to do as part of its mission to its consumers. Other main stays in, Fast Fashion include Uniqlo, GAP, Primark, and the TopShop. Arguable, the price point offered by these brands are already cheaper compared to the designer brands of which they try to mimic the design. Yet there is even lower tier that offers a lower price point. This shall include, Misguided, Forever21, Zaful, Boohoo, and Fashion Nova.

1.3 Fast Fashion Business Model

As already mentioned previously, speed and quantity make up this business model. Zara as a pioneer, enables this as they work on a 15 day cycle for a garment to go from design to distribution. Fundamental strategies that enable speed and quantity to be realized is the capacity to stock less merchandise while exhibiting collections often. Since its conception from the city of Coruna in Spain, the brand has now wide reach of demographics and geographics, present in all age groups across all continents in the world, respectively. It boasts a distribution network of 2000 retail outlets in 88 markets which are strategically located in shopping districts of key cities of every country. Normally, fashion houses participate in fashion shows twice a year. Yet for Zara, the several collections are displayed each season, revolutionizing the world of fashion.

Above all these high-level strategies in place, is a process pioneered by the brand. Zara as a brand, has hired 200 designers located in the company's central headquarters in Spain to gather information from trends to customer feedback and turn that to design and execute it to merchandising and distribution. Getting even closer to the information, the brand probes on emerging trends by scouting the streets and malls where they are present. The headquarters in Spain also houses the production and logistics of Zara which is an advantage for the fashion house to remain flexible and agile as the changing landscape would demand.

The brand's marketing, as guided by their business model, lures its customers to purchase often and quick. The former as exhibits is displayed every four weeks, and the latter as limited merchandise or stocks are available per item. Consumers fall into this belief and rightfully so, feel the need to grab the chance to own the latest find at their favorite Zara store.

Zara's strategy is unmatched when it comes to timing and efficiency. The brand's products are manufactured to sell quickly, and closely follow luxury street wear fashion trends. The behavior of Zara consumers reflects their fast fashion strategy, as they are aware that if they like an item, they must purchase it right away, because it will likely sell quickly or be replaced by new stock. A key player in Zara's business model is that they have limited production seasons. This leads to extremely high turnover of products. Specifically, product inventory changes every 15 days. A massive expense for many companies is the storage and inventory costs of keeping products in one place to be later brought to shelves. Zara, among other fast fashion giants, saves on these expenses as a result of their quick turnover and the minimal time that clothing spends in storage. Business risk is reduced due to the brand's in-depth knowledge of consumer behavior. In other words, Zara knows what their target market wants, and they don't produce items that are risky or won't sell. A few recent sales figures represent Zara's continued success. In 2014, Zara's sales reached 11.594 billion euros, an increase of 7% compared to the previous year. What is extremely interesting about Zara and other fast fashion brands, is that they save money on advertisement and rely solely on their large retail outlets in high-traffic locations, as well as consumer word-of-mouth.

Zara is just one example of a key player in the fast fashion industry. Many businesses like Zara, are shaping the modern consumer's relationship with clothing and fashion. Imagine if we could reframe this way of thinking, and instead of the consumption of fast fashion, we replace it with the consumption of upcycled, local and alternative materials. The value that consumers see in fast fashion is that it is constantly new and trending. What if this value was replaced by the value of uncycled deadstock, vintage, second-hand clothing, local textiles, and alternative materials all together? Luckily, this is not too far-fetched. We are gradually seeing a shift in the way that consumers view fast fashion. In order to better understand the detrimental impacts of fast fashion as consumers become ethically conscious. Before we get ahead of ourselves and look into brands that are emerging in the alternative materials sector, we must consider the textiles that are most harmful in the industry and why they are harming our people and our planet.[2]

1.4 Global Textiles

So why is fashion relevant to the discussion on local and global textiles? Fast fashion materials have a massive detrimental global impact on the environment and human

---

[2] Wharton Global Youth [2].

health. Textile production has become an intricate system of supply chain partnerships and networks around the world.[3] Producing this volume of clothing items at
such a fast pace involves the use of non-biodegradable synthetic materials, large
consumption of water, and massive amounts of waste.

It is no secret that many of the materials that are used in fast fashion clothing products are harmful to the environment. The overuse of synthetic fibers and polyester
have the largest impact on the environment. When produced, polyester emits a significant amount of carbon dioxide. It is considerably more harmful than other materials
used to produce clothing. Furthermore, polyester takes years to degrade. Unfortunately, many global fashion brands use these materials in their clothing because it is
both inexpensive and easy to produce in large volumes. Since the year 2000, there
has been an increase in polyester production in the twenty-first century. This increase
should motivate companies to choose sustainable fabrics in order to decrease their
carbon footprint, yet companies continue to produce these materials in large volumes
due to the ease of manufacturing and producing polyester.[4] In addition, the negative impact that these materials have on the environment should be considered by
consumers as they choose which brands to support.

When consumers and the general public think about the impact of the clothing
industry on the environment, we often fail to consider the entire process. When
producing toxic materials, the amount of water that is used to run factories and
clean products is immense. For example, 'it takes 10,000 L of water to produce
one kilogram of cotton or approximately 3,000 L of water for one cotton shirt.' In
addition to cotton production, the coloring of clothing is toxic to the environment
due to the chemicals that are emitted into the ocean upon disposal.[5] Another thing
that consumers and companies overlook when it comes to clothing production is the
location with which these products are produced. Unfortunately, many companies
will build factories in third-world countries that have limited environmental regulations. Brands will do this on purpose so that they can generate the highest turnover
for the lowest amount of initial investment with regards to their two most valuable
resources—time and money. In addition, they have increased access to disposing
chemicals into the ocean. Regrettably, the wastewater created is extremely toxic and,
in many cases, cannot be treated to become safe again. On the other hand, producing
clothing in continents such as Europe or North America is often more expensive
because they are obligated to produce materials in regulated conditions and are fined
for breaking any principles in place within these regions.

So why is this relevant to the overall discussion of the localization of textiles? The
rate at which we are consuming and discarding textiles is contributing to mass waste
not only in our own landfills but also in developing countries. Since the increase
in globalization following the industrial revolution, the access that companies must
developing countries has increased and become detrimental to the overall well-being
of workers and the state of landfills. More specifically, 6 percent of textiles produced

---

[3] UNEP [3].

[4] Common Objective [4].

[5] WHO [5].

each year are unpurchased and end up back in the landfills of developing countries. 'While 80 percent of the clothes discarded by consumers in major cities could be reused, these typically end up in landfills too. In the US alone, clothing landfills occupy more than 125 million cubic yards each year and the worst part is that most of these clothes are made from non-biodegradable materials.' What if we were to produce, consume and dispose of clothing responsibly? Why can't we localize this entire process and create a circular economy system in each respective country? Unfortunately, rethinking the ways in which clothing has been produced for decades is easier said than done, especially when companies are able to save time and money by continuing with unethical practices that they have become accustomed to, and that consumers continue to support. With that being said, it is absolutely imperative that we shift the priority from money to the overall state of the environment as we continue producing and distributing clothing around the world. The clothing choices we make do not just affect us personally. They also have global implications on the environment and the livelihoods of millions of people in developing economies. If we all choose a slow fashion way of life by reducing our consumption, we can alleviate these detrimental impacts. What is important to consider, is the fast fashion presence in specific countries, in contrast to what these countries are doing to address fast fashion and shift to local textiles. The fashion industry has caused a substantial amount of damage to our environment. However, if we start to take proactive steps toward advocating for a green-friendly fashion industry and becoming an environmentally conscious consumer, we can finally slow down climate change. The fashion world was not always like this, and in many cultures, it still isn't. Thousands of cultures worldwide have continued to produce unique, eye-catching textiles ethically and locally. These cultures and countries can be used as a benchmark for larger companies and set precedent for the ways in which they place value on tradition and living in harmony with the environment.

1.5 African Fabrics for Fashion

Typically, west African fashion textiles are used by indie fashion brands. Enthusiasts can argue that the African fabric suits any kind of fashion ensemble. These fabrics find its use into most type of alternative fashion. For one, 'Mitext holland' caters to a range of traditional patterns in a variety of earthy color pallets that potentially adds to the retro-style of the garment. On the other hand, 'Getzner' offers a wide -spread of more low key yet still stylish patterns for a creative alternative suits. Finally, 'Vlisco' which comes with its best bet on colorful patterns, also produces a classic black and white fabric which showcases a brilliant feature of angular and visually pleasing patterns. Simply put but intricately curated, the world of gothic fashion, might be the obvious clientele for these mentioned examples.[6]

---

[6] UFO Themes [6].

In addition, as cited in UFO Themes (2021), another exhibit was held in London which is referred to as 'Fashion Cities Africa' and which was dubbed as the 'first major UK exhibition dedicated to contemporary African fashion', gathers designers across the continent. Key cities in the region, however, stood out with the most number of participating designers namely, Johannesburg in South Africa, Nairobi in Kenya, Lagos in Nigeria, and Casablanca in Morocco. The exhibition celebrated the region's diverse and emerging fashion scene. Helen Mears, one of the museum's keepers of world art explains, 'We want to reveal the diversity that exists across the continent—and within single cities—and show that wax print is only part of the story of African fashion. Each of the cities featured has it own fashion scene: in some cases emergent, in others more established.'[7]

It could be derived from this, what also is happening in actual, where African designers are being noticed as major players in international fashion while others emerging names keep diversity and variety at play between the global fashion and local heritage.

The exhibit in fact made waves in the online space particularly in the different social media platforms. People have talked about clothes made by influential designers in the industry as well as some breakthrough up and coming ones. One example is the 'The Sartists', a Johannesburg-based collective that displayed a range of urban and a post-apartheid wear that made it big through the brand's Instagram account. According to the brand, they aim to 'recreate and communicate African authentic stories through style'. Another stand out was the Lagos-born designer, 'Amaka Osakwe' who displayed a wide range of West African textiles and fabrics. The designer's brand exhibited her women's line label 'Maki Oh'. Featured dresses had caught the attention and interest of Beyonce, Rihanna, and Michelle Obama. What was a highlight for the brand was the artful combination of traditional fabrics and textures with a tasteful contemporary twist to it. On the contrary, Osakwe's signature is the heavy use of the traditional textile art process 'Adire'—a form of hand dyeing.[8] To keep the dialog memorable to the visitors, the exhibit shares the traditional methods that were done in the past while combining a new perspective to the methodology.

These exciting events, like the previously mentioned, raises awareness and the broadcasts profile of modern African fashion. It sheds light as well to a different image toward the region while attracting new public perception toward the rich and diverse styles available in this medium. These kinds of avenues allow the continuation of conversations started by the ones who created such possibilities and to those re-creating them at present times. It is vital that such effort is made for the preservation and innovation of this art form. The new audience will have a taste of the color and vibrancy of the African fashion set in the times that is more curate to fit the relevant times. Equally, promoting the legacy of these fabrics and textures.

---

[7] Situating Alternative Textiles in Kenya by Fashion Revolution—Issuu [7].

[8] HelloBeautiful [8].

1.6   African Fashion Shows Stirs Global Impact

As the world continues evolve, we find ourselves saying more frequently, that the 'world is small'. And this is realized too in the way fashion operates. From fast fashion to emerging fashion shows, Africa's take on it is heard even louder. As the global stage centers to this region of the world now. However, as the world we know is getting more and more accessible from west to east, one does not have to travel to Africa per se to experience African fashion. During the previous decade, the number of exhibits and fashion shows catering to African designers and the likes have increased. According to the same source at UFO Themes (2021), influencers in the industry predicts that time for Africa to shine has come.[9]

As fashion capitals would have it, most new shows are based in the likes of New York City, where the annual African Fashion Week was initially launched back in 2009. This is considered such a feat or a milestone as it provided an avenue once again to raise awareness about the subject—African fashion and the professionals who run it. No surprise, that London, another fashion capital, had it going as well. Two years after the New York launch, came the annual African Fashion Week in London celebrating the city's rich cultural heritage and the mix of African and Western Fashion. In addition, as it has somehow become a widespread trend, other countries have followed pursuit, hosting their version of the annual event. Cities include Holland, Ethiopia, France, Australia, and Belgium.

1.7   Africa's Found Spotlight

According to Atim Oton, a Nigerian-born designer, exhibits such as fashion shows are opportunities for African designers to establish a credible and sustainable business that would lead to a revolutionary change for Africa. Oton describes the increasing global exposure as 'fabulous' for the continent aides for new sets of eyes to perceive the region differently and as such. Additionally, she believes that the rise in popularity will bring about an important shift for African designs leading from traditional one-off trends to actual contributors to the global fashion industry. Moreover, Oton describes this time as t 'Africa's time to shine'.

In 2000, Fati Asibelua created her fashion label called, 'Momo' which later was launched her namesake line, 'Asibelu'. The brand was admired by her patrons for the tasteful and unique prints of textile designs mixed with the infamous traditional African patterns with modern shapes. Her creations were found to be deeply rooted in African culture yet blended with modern style. This being the case, Fati Asibelua is one out of many emerging designer to hold the torch to a fashion revolution in the African fashion landscape.

Another designer that became top picks by A-listers in Hollywood, is a South African designer Malcolm Kluk and Christian Gabriel Du Toit who launched

---

[9] UFO Themes [9].

KLuK FGTD in 2000. The client list includes Beyond, Rachel Weisz, and Charlize Theron. The brand's collection ranged from featured bespoke bridal to ready-to-wear ensembles.

As these ongoing trends and milestones are made by these designers and across key cities across the globe, new and upcoming African designers continue to innovate this craft and the use of these textiles. Heads turn toward Africa, and these new designers get the same chance if not better chances to branch out. Atim Oton concludes with hope and pride that the global reach for African designers and their work can only get better from here as fashion shows and exhibits continuously share the story and culture of new found love for African art in fashion.

2   Sheding Light to a Region and Its Tradition

Having said all the above, one can start wondering to dig deeper, why Use African Fabrics? West African fabrics consist of bright and visually engaging patterns and colors like today's independent fashion trends. The rise of indie outfits featuring patterned textiles with vibrant contrasting colors call for a demand for West African designs and fabric. Additionally, it's worth noting that independent clothing designers are looking into creating a carefree and energetic aesthetic with bright hues and cheerful patterns. Fabrics from Vlisco consist of vibrant hues and modern patterns. Thus, in the world of indie fashion these fabrics would be even more useful than any other West African textile.

On the other hand, the celebrated Hell's Belle's show by Mumbai-born designer Manish Arora was not only influenced by Americana, but also African fashion. Although his women's collection was themed Wild West, he made use of African wax prints in all his designs wherever possible. The newly awarded Legion d'Honneur was widely celebrated for his brilliant combination of colors and textiles, particularly his attention-grabbing use of wax prints on denim jackets and skirts. As the first Indian creative to be rewarded for fashion and arts in France, according to Arora, it took him a while to believe that the French institution recognized his truly deserving work of art.

2.1   *Combining West African Fabrics with British Counterculture*

Empire Textiles are keen on combining British alternate native fashion with West African textiles believing that this would create some of the most unique and fascinating garments all over the world. As British fashion is known for its wide range of alternative options from rough-and-ready outfits to highly customized suits, the subcultures of the country have also produced quirky, macabre fashion trends. On the other hand, West Africa is renowned for its vibrant and intricate designs. To merge these two cultural fashion trends into one would undoubtedly yield an exceptional collection of garments.

2.2 Stories and Meanings Woven into African Wax Prints

As history would suggest, the batik method used in Africa, was brought by the Dutch via Indonesia. However, do we know the meaning behind the emblems that we see on the patterns? African wax prints tell a colorful narrative that has woven history and culture throughout time. As the saying goes, every picture tells a story. How fascinating it is to explore the rich culture and story behind the bold prints and dazzling color in the rolls of handcrafted textiles.

Take the imagery on Wax Prints as an example where each image would mean something else compared to others. Interpreted by 'Stories and Meanings Woven into African Wax Prints' the below explains a few designs commonly used on fabrics, 'You fly, I fly' referring to an image of a bird flying from a cage known to be a staple to newly-weds. One interpretation of this image is to symbolize an expatriation from the home origin, another interpretation is believed to be a warning code to the new husband. Vinyl disc appearing on fabric is later referred to as 'Gramophone'. May sound sentimental for many, as the vibrant circular design testify a deep appreciation and taste for music dating back to the original record players. Tortoises as referred to as ancient reptiles exemplify qualities of patience and resilience. Often understood as a dense and complex signifier, as when viewed from the top the shell draws whorls and indentations.

Birds in flight, represent money is not managed well and may simply just fly away from its master. On the contrary, diamonds would suggest more obvious image representation of money on wax prints. Diamonds being native as well to the region is a viable representation of the local culture. A stick design symbolizes a humorous message, the 'Stool' is a good example for that. According to the same source, "*The idea behind this is to take a seat and sit down if you have anything meaningful to say. Anything worth a discussion deserves the time and effort to draw up a stool to sit and chat.*" A popular design is the Spirals of dots forming a circle. The dots or circular shapes depict the water droplets while the spiral formation mimics the effect when the water is disturbed by a foreign object. The water's ripple effect is illustrated in the spiral formation. Water as valued by many brings forth life and it is thus obvious that its key importance implies having confidence and good health. Such is Life. Last but the last among the designs and its meaning, is a design that is featuring guinea fowl. With this design, however, the wearer must be sensitive to wear such for special events or occasions as it comes with a level of expectation as the animal is perceived with great delicacy and is culturally enjoyed in several dishes.[10]

---

[10] UFO Themes [10].

Aside from the Wax print, another iconic African fabric that took the world over, is *Angelina*. Inspired by the Ethiopian clothing tradition, 'Angelina' has this V-cut shaped ornate main pattern with an embellishment of a dotted band on the edge. Interestingly, the name was derived from a very popular song by a band from Ghana, which was also released during the time the fabrics were launched for sale. The beauty of this fabric is that it comes as well in different color combinations, a rich selection to say the least and is almost always worn in dashiki style. To define further, the dashiki is a type of pull over outfit either as a shirt or a dress with yet again the V-cut shaped pattern forming the collar at the front while opposite to it is the pattern.

There are few other designs that were named after public figures. We start with Nkrumah's pencils believed [the pattern] to celebrate the intellect of the legendary pan Africanist.[11]

Another design is illustrated in a cauliflower like shape where a cluster of three trees are bunched together in an almost brain-like shape. This is called 'Kofi Annan's brains'. The name is coined for its un-usual pattern design which was introduced around the same time Mr. Annan was about to end his second term as UN—Secretary General. Such design was there to honor the brilliance of the Mr. Annan A consequence of historic win in US politics in 2008, former US president Barack Obama was honored by consumers along with his wife, then-first lady Michelle Obama. Two of the fabric patterns that were introduced to the market almost at the same time were called 'Barack Obama's heart' and the other was 'Michelle Obama's handbags. Following this feat, soon as Mrs. Obama has set foot in the region, with her series of travels to the region. In fact, as the consumers primarily dictate the trends while brands are there to try to influence or supplement it, none of the names over these patterns came from the company but from the consumers themselves. Where fashion finds relevance to history and where brands supplement this by celebrating personalities onto their design, that is what the fashion revolution in Africa is coming about. Speaking about revolution, as diverse as the region, these names can be easily referred to differently by neighboring African countries. For instance, Angelina is referred to as 'Ya Mado' in the Democratic Republic of the Congo (DRC) simply because of a video where dancers were seen to have sported the same fabric pattern.[12] Similarly in Ghana, 'Senchi Bridge' is what is referred to for the same pattern called 'Cha Cha Cha'. This pattern evokes the legendary Congolese rumba tune entitles, 'Independence Cha Cha' in the 1960s. Yet for the people of Ghana, the 'Senchi Bridge' depicts the suspension of this bridge on the Volta River which was believed to swing widely when traveled on. And if that is special, the African neighbor country, Togo refers to the same pattern differently as well. For them, it's the back of the chameleon.

Be it in Democratic Republic of the Congo, Ghana, or Togo, the labels put on the pattern reflects the stories of people referring to it. It mirrors the social norms at that time. For a long period, women take up the bigger pie of consumers of fabric for perhaps obvious creative reasons found on these textiles but also because of the need

---

[11] Africa Renewal [11].
[12] Africa Renewal [11].

for cloth allocation—a two to three piece ensemble which include a top, a bottom, and a waist cloth. While men would have a simple shirt out of the fabric to make up their ensemble partnered with a pair of trousers. Frankly speaking, it does not make much difference to the modern style outside Africa. Now with a modern take to it, Johannesburg-based fashion designer Tanya Kagnaguine, coins a term, 'Afrochic' which means a good mix of traditional design and modern fashion wear.

By now, we have established that from all the new terminologies coined or used to refer to fabric or trend evolution, this document has framed the conversation around the African fabric where daily lives in the region have clearly been the inspiration, supplemented with the African culture and also an extension of values and social norms.

2.3   An Appreciation to Vilsco

On top of the previous exhibits already mentioned above, what was interesting is that an earlier exhibit was made to show a particular highlight to Vlisco. Entitled, 'Vlisco: African Fashion on a Global Stage', the exhibition was launched at Philadelphia Museum of Art.' The bigger exhibit referred to as 'Creative Africa' combined innovation and tradition to put the coming of African art and design to the global stage.

Specific at the Joan Spain Gallery, dubbed 'Vlisco: African Fashion ona Global Stage' explored the history of the Dutch Company and shared how brands in the textile industry managed to become influential and not just in specific areas in the continent but throughout the entire fashion industry worldwide. As hot is known by many, the bright and bold patterned fabrics were front and center as spectators discover the exhibition where they were given a snippet of the company's effortless blend of tradition and luxury where simply put designers from two continents, Europe and Africa have managed to find a common ground in interpreting the beauty of West African fabrics.[13]

Interestingly, Vlisco as a brand is quiet global with regards to how they produce. To start Vlisco's wax printed textiles were both designed and manufactured from three locations, Europe, China, and Inia while the most sought after luxurious fabrics are specifically produced in the Netherlands.

2.4   From West to East

To momentarily close a chapter full of conversations found buzzing around the US, an African designer was invited to dress the White House. By now, it is no secret that the former First Lady, Michelle Obama is a fan or a supporter of African designers as

---

[13] UFO Themes [6].

she was frequently spotted wearing African classics like wax prints and Swiss Voile to public appearances. Enthusiasts were quick to appreciate such sightings on the former first lady as then the African style and material were given a modern look and proven that even[1] traditional print and fabrics can dress any occasion. So much so that, Nigerian designer Duro Oluwo was fortunate to be selected among other designers to display their work in one of the rooms at the White house. This project has been an ongoing tradition in the US for the President's wife to invite designers to put their own take on a room each and under the Obama administration. Fortunately, so, as we speak of global scale, during the said administration, the selection of designers included talented individuals from US, Europe, Asia, and of course, Africa. All of which to part of a historic first for the world's most influential building.

Meantime, strengthening relations among other countries or markets is key to keep the momentum of growth for the African fabric. Take it for the Turkish president who pledge to improve trade relation for West African textiles during his visit to the region. And sole purpose of this visit was then to pave way for future of trade which is vital to keep the network of African fabric development happening on this side of the world happening. This only is not very promising time for the entire West African textile industry, but also a momentous event for the region, where for the first time a Turkish President visits the Ivory Coast and Guinea. In 2014, data showing the trade between these two nations, Turkey and Nigeria, reached USD 2.4 billion. In addition, included as a top priority for the Turkey-Nigeria Chamber of Commerce is the strengthening of importation particular to clothing materials. In the future, investments made by the Turkish government to Nigerian textiles will only fuel more growth and expand trade for the infamous African textile.

In celebration of this feat of cultural fusion and solidarity another opportunity that highlighted the fabrics has taken place at the UN Educational, Scientific and Cultural Organization (UNESCO) headquarters in Paris, France. One of the featured materials shown was a classic turned into a digital animation was the Jumping Horse, also referred to in French as, "*Je cours plus vite que ma rivale*" which translated as "*I run faster than my rival*".

The year 1995 was a remarkable year as it was leading to the Beijing women's conference. Which happened also during the time where women's rights were debated and fought for alongside conversations on the legal statue of polygamy. This happened to be the same year where West African design pattern came out. In fact, a design pattern was linked to these conversations happening globally. Dubbed as, '*Si tu sors, je sors*' which meant, '*You leave, I leave*'. What made this pattern relevant was the idea that depicted the need to respect the rights of the young women of the region. Many if not most, would simply look at this as a spousal warning again infidelity, yet became a cause of being for the young generation of the nation. This was important to note, as going back to the new base of consumers continuously seek for brands to support or in this case, simply fabrics that will reflect their values. Esi, having been able to curate a video mapping of these fabric, believes that she was able to raise a level of awareness of the social relevance of the fabric encapturing innovation, creativity, and yet again consumer's choice to share and celebrate life through clothing.

'This is us, these are our stories", she said.

Finally, the connection between the future of fashion and the rising popularity of African prints showcases the talented designers from the region that projects a sustainable business and interesting art form to appreciate. One may come first then the other, but that's what remains interesting for fashion and these textiles, as there is a continuous innovation on how brands can play on both the story telling and the business potential out of it. To cite this, LVMH fashion 2014 finalist, Amaka Osakwe of Maki Oh proves that there is a continuous curiosity and recognition among the Black Fashion Designers and the work they do. With the efforts of putting 'sustainability' at the forefront of all discussions and the relevance of African fabric, it does make sense that more conscious fashion industry is on its way. Awareness drives further curiosity and that effort coming already from the public or the consumer itself, brands role now is to respond than influence. Different sectors in society do enable this effort of providing 'Conscious Fashion' alternatives if not an entire switch from traditional fashion we currently have. The strength of Maki Oh's brand is that not only is an addition to its brand's value that guide them to produce more eco-conscious collection, but it is in fact at the core of what they do. This is a strength that perhaps gave her an edge in the said competition, as now African designed patterns find even more relevance to the conscious new generation of buyers. Truly, it is a cycle that benefits many of the stake holders in and out of this growing trend.

> "Styles and trends and preferences go back and forth all the time especially today what we call popular fashion, it may be influenced by Western fashion but we also are seeing African designers who are drawing from Western traditions but also from local styles," Quick added. "So there's this very fluid exchange of ideas, trends and preferences. I think there is such an appetite for African fashion today, much more so than 10 years ago," concluded Quick[14]

2.5   Fast Fashion and the Philippines

Asia has always been the capital of the world's textile production and arguably the source of fashion trends. For many centuries this has been the case due to factors that range from increasing population, rising middle class, and cheaper labor compared to other economies. Yet India stands out as a real game changer as it produced the largest share of textile production. Many thanks to its culture of vibrant colors and festive affairs, Indian manufacturers had developed throughout time the most sophisticated method of playing around with cotton with regards to its known quality of lightness, breathable and long-lasting dyes. All of which gives birth to the fabric's dazzling colors. In fact, back in the nineteenth century, Indian textiles were in high demand that producers were quick to create special lines for export to Southeast Asia, Africa, and Europe. This enabled trades with adaptability and a level of flexibility given the demand in these markets.

---

[14] UFO Themes [6].

Today, fashion shows are housed in key cities across two major continents, Europe, and North America. Notable cities include New York, London, and Milan who set trends and launch new and upcoming designers. However, Asia's strength in textile design and production remains to be admired and at some point, exploited by the world. China and India being the leaders in this regard. The split of design and manufacture between the West and the East is rather new in the history of trade or fashion in retrospect. Such development reveals how Asia's textiles supplement that of the modern fashion industry, if not influencing a portion of the style as well. While the Western's appropriation of those aesthetics demonstrated further to the rise of fast fashion that consumers experience today.

Shedding light on the fast fashion industry in the Philippines is worthwhile given the traditional local textiles that are going global from this region. It remains the same as any other key cities that global fast fashion houses dominate this country. Zara was first to enter in 2005, then Forever 21 in 2010, and Uniqlo in 2012. These are the brands that slowly found the business sense to be present in the Philippines. They have taken place in shopping centers and have made their presence felt in retail stores. Traditionally, Filipinos enjoy shopping at 'tiangge'—a local term for bazaar. But since the items sold at these spaces are slow to update and offer convenience to its shoppers, eventually modern facilities in big shopping centers became the new normal of shopping.[15]

Like other consumer behavior, 65% of Filipino adults were reported to have thrown away clothes in 2016, according to the 2017 YouGov Omnibus survey. As fashion is known as a pollutant from the way it is produced with the use of chemicals to dye to form shapes and textures of the fabric, so as they are harmful when these fabric wastes fill up dumpsites and take years to decompose. When developing countries like the Philippines suffer from natural calamities it is easily affected by the aftermath of it. As it takes 2700 L of water to make one cotton shirt equivalent to one person's liquid intake for 3 years, water resource becomes so valuable that developing countries like the Philippines feel the effect of drought during the dry season, almost immediately compared to other neighboring Asian countries.

Traditional textiles that originate in the Philippines such as pina fabric and abaca fibers, are locally sourced and highly sustainable alternatives in fashion. As we approach 2022, we are seeing these local materials re- emerge in Filipino fashion and the global market as companies become more aware that they must enter the alternative material era of fashion. Before examining the Philippines' local textiles specifically, let's look at a company that is not only embracing alternative materials, but utilizing otherwise wasted deadstock to create luxury clothing.[16]

Thanks to the fashion entrepreneurs, the country of the Philippine's weaving heritage finds a renewed spotlight. From the onset, the use of this traditional method provides opportunities for tradition to provide sustainable source of livelihood while inflicting a sense of pride to those who find interest in these products. In addition, there is an annual celebration of local fabric that commences during the beginning

---

[15] Perez [12].
[16] UFO Themes [6].

of the year as the country hailed the month of January as Philippine Tropical Fabrics Month. Truly a spectacle on its own, visitors are delighted to the development in the Philippine Fashion landscape. As there are forthcoming trends where innovative use of traditional fabrics is showcased and championed by Filipino themselves. Fortunately, with the availability of modern tools for weaving, these local textiles find a new light and form, where stories from ancient past are shared to the new generation.

2.6   Respecting the Ancient Craft

It has to be considered that these products inspired by the stories from indigenous textiles are both cultural art forms and national treasures. It is therefore a responsibility or a mere sensitivity from the owner to seek knowledge about these stories, values, and disciplines woven into the textiles. And similarly, from the buyer to educate in the process of creating these products to selling them. These artisanal fabrics are rooted from rituals and traditions and must be used ethically.

Vela Manila, a Philippine accessory brand explains the Yakan tribe-designs draw inspiration from the natural landscape and surrounding area in the province of Basilan. The fabrics as she continues are characterized by intricate geometric patterns with vivid hues which are sourced from pineapple and abaca fibers. These are textiles dyed with herbal extract and used as part of the entire product. She concludes that the entire process may take a week as it is labor extensive to weave just one meter of the fabric alone.[17]

Apart from the natural material, for other tribes in the country, the color of the textile holds significance. The Maranaos view the color yellow with high regards to royalty or those who hold high status in life. As Abdullah T. Madale writes in 'Textiles in the Maranao Torogan', a commoner who wore yellow could be ostracized or even beheaded. Madale continues to explain, that wearing green can omit or project humility of one's accomplishments, which can also stand for peace and solidarity. While Red as normally for bravery and violence, white stands for mourning and finally black for silent dignity and purity.[18] Truly, color is more than just an aesthetic pleaser.

By now, it may seem that these inspirations are complicated as they are deemed holy. Partly true as there are hidden meanings to each stroke and color choice. The last thing that a user would not want to be referred to is disrespect to their other's beliefs. Research goes a long way, yet another easy way to avoid this kind of misuse is to patronize brands who truly understand the stories and principles behind the designs they use. It is easier to think that locals from those areas are more affiliated to the traditions better than those who are from elsewhere. In the end, it is a beautiful art form and just like that, it pays to know the inspiration from which it draws. Today's modern technology allows a wearer to gain information regarding indigenous textiles

---

[17] Manila Bulletin [17].

[18] Esquiremag.ph [13].

that they may have just bought or simply possess and more importantly know how to wear them appropriately with a simple tap on their mobile screens. There's a lot of cautiousness that must be put into wearing these artisanal works with pride. And educating is as important as the creation process.

2.7 Give Proper Credits

It is only appropriate that recognition be given to the people who dedicated long hours of labor from conceptualizing to producing the textiles we see today. These intrinsically designed fabrics hold meaning and significance to its roots and entertains its new audience. All components that go into the design are equally important—stories, creativity, and labor.

From the distant past to the new century, one of the finest arts was born in the town of Taal, Batangas. Considered as refined art in Philippine society, the tradition of intricate embroidery is considered as a refined art because of its complex and well-embossed appearance.

During the Spanish times, one of the most expensive and rare fabrics used to create a Barong Tagalog was the pina. It is traditionally used in embroidery because of its soft, flexible, and durable qualities. The fine, off-white threads of Pina is taken from the mature parts of the plant, manila the leaves of a Spanish variety of pineapple.

Traditional embroidery in the Philippines has withstood the test of time. Today, embroidery is used to embellish and enrich handcrafted accessories such as bags, clutches, hats, and other native products. For instance, high quality Philippine products such as abaca and raffia bags are adorned with intricate embroidery patterns to increase the aesthetic value of these indigenous handicrafts. It has such a remarkable form of art in the Philippines that many associated products are being exported in almost all parts of the world.

2.8 Local Textiles

2.8.1 Pina

A popular alternative material being integrated into fashion in the Philippines is Pina, a fabric made from the fibers of the leaves of the Pineapple. The future of the fabric is obtained by cutting the leaf from the Pineapple plant. The fiber is then derived from the leaf. In fact, the leaf fibers are mostly long and stiff. Each of the strand of the produced fiber is scraped by hand and is then woven or knotted together to form a continuous filament which will later be hand woven and made into the proposed textile fabric.[19]

---

[19] Pineapple Industries [14].

This fabric is quite famous in the Philippines. The material is apt for the weather and the plant from which it comes from can easily be grown and produced in Philippine soil. Artisans have found value to this delicate fiber, considerably one of the most delicate in texture than other vegetable fibers out there. A kilo of leaves may provide up to 15–18 pieces of white and lustrous silk fiber about 60 cm long while retaining dyes.

Pina cultivation goes back to colonization of the Philippines, though is grown in other tropical regions as well. Using pineapple leaves as fiber is an excellent eco-friendly alternative to leather and synthetics. The fiber silk was even considered the queen of Philippine fabric choice. Although pina was a valuable resource worldwide, the cheap production of cotton emerged, and pina textiles became less relevant in the fashion industry. In recent years, the material has been revived due to the demand for natural, alternative resources. Particularly in luxury fashion, pina is commonly used as a leather alternative. The older that clothing produced with pina grows, the softer the material feels and looks. This poses a huge opportunity for the luxury industry, given that the material is timeless, soft to the touch, and a beautiful beige color. There are many ways to use pina in the production of textiles and its versatility and beauty will be valuable moving forward in the sustainable fashion industry.

2.8.2 Abaca Fibers

This type of fiber is extracted from the leaf that grows around the base of the trunk of the abaca plant. Abaca is a close relative of the banana plant that is native to the Philippines or other humid tropical countries. To harvest the abaca plant is laborious as each stalk is cut into strips which are then scraped to remove the pulp. The fibers are washed and dried naturally. This material is another leaf fiber that is composed of long slim cells. Abaca is valued by artisans for the fibers' ability to hold strength for structure and to resist water damage for longevity. Total fiber length could extend up to 3 m. The best type of Abaca gives a lustrous, light beige color and a strong one.[20]

A proud and successful eco-conscious material user and Filipino fashion designer, Dita Sandico-Ong, shares her success in the fashion scene. She had the opportunity to utilize natural fibers form banana, abaca, pina that are all native from the Philippines. With her creativity and the natural ability of these fibers to elevate those designs, her work was featured in the fashion capitals of the world, from West to East.

2.8.3 Inabel

This fiber hails also from the Philippines, in the northern province of Ilocos. 'Abel' from its root word is the local dialect's terminology for weaving, while 'inabel' refers

---

[20] Esquiremag.ph [13].

to woven fabric. In the weaving industry, locally speaking, these terms are known to refer to the fabric native to the province uniquely.[21]

Inabel is a cotton like fabric that could be plain or with pattern. Similar to the other natural fiber, abel cloth is appreciated for its lightness and strength. Interestingly, these natural fibers are viable options for luxury brands as they turn their business models to more eco-friendly. Take Filipino designer Patis Tesoro who had the chance to meet with Salvatore Ferragamo, Valentino, and Giorgio Armani who have shared their interest toward these natural fibers.

3 Conclusion Deadstock and Reformation

So, if fast fashion and the global textile industry are harmful, how will designers in the fashion industry build and maintain businesses that source local textiles?

Christopher Reaburn, a British fashion designer, turned surplus military garments into eight garments for his first show at London Fashion Week. This statement exemplifies sustainability and waste management with boldness and innovation. In the decade since this show, Raeburn has built his brand around surplus fabric and clothing, including military jackets purchased in bulk from the UK Ministry of Defense.[22]

An increasing number of designers are turning wasted and found fabrics into business. Contemporary fashion-school students are discovering their ability to 'scour flea markets and warehouses for discarded textiles can be scaled into global brands.' Another example is Reformation, a sustainable women's clothing line, who started outsourcing deadstock from Los Angeles-area stores and factories. Since launching, the brand has built a $100 million-plus value, while online fast fashion giants incorporate vintage and used materials into 'Reclaimed Vintage' lines. Today, the company approaches factories it works with around the world about whether they have any interesting fabric left over from past jobs.

Moving forward, it will be crucial for global fashion leaders to lead by example and maintain the momentum that the younger generation of leaders has established. By researching authentic, sustainable, and influential brands and designers like the ones mentioned in this chapter, global brands can begin to innovate and adopt the methodologies of local brands. The time is now for change. We no longer have the resources to continue approaching fashion production on such a detrimental, global scale. Leaders must not only focus on how they can maintain profitability, but more importantly, address the neglected pillars of sustainability—social, and environmental—equally. This will include preserving authentic designs, representing different cultures appropriately, and shifting from unethical global production to a more local and sustainable approach to the manufacturing and production of clothing.

---

[21] Narra Studio [15].
[22] The Business of Fashion [16].

References

1. Bick R, Halsey E, Ekenga CC (2018) The global environmental injustice of fast fashion. Environ Health [online] 17(1). Available at: https://doi.org/10.1186/s12940-018-0433-7
2. Wharton Global Youth Program (2022) Zara's 'fast fashion' business model—Wharton Global Youth Program. [ONLINE], available at: https://globalyouth.wharton.upenn.edu/articles/zaras-fast-fashion-business-model/. Accessed 15 Oct 2022
3. UNEP (2022). Putting the brakes on fast fashion. [ONLINE], available at: https://www.unep.org/news-and-stories/story/putting-brakes-fast-fashion. Accessed 15 Oct 2022
4. Common Objective (2022) Reducing fashion industry water usage | common objective. [ONLINE], available at: https://www.commonobjective.co/article/the-issues-water. Accessed 15 Oct 2022
5. WHO (2022) Water pollution control—a guide to the use of water quality management principles. [ONLINE], available at: https://www.who.int/water_sanitation_health/resourcesquality/watpolcontrol.pdf. Accessed 15 Oct 2022
6. UFO Themes (2021) West African textiles take centre stage in Philadelphia—Empire BlogEmpire Textiles Blog. [ONLINE], available at: https://www.empiretextiles.com/blog/portfolio/west-african-textiles-take-centre-stage-philadelphia-usa/. Accessed 15 Oct 2021
7. Situating Alternative Textiles in Kenya by Fashion Revolution—Issuu. [ONLINE], available at: https://issuu.com/fashionrevolution/docs/fashrevke_policyreport-sept. Accessed 15 Oct 2021
8. HelloBeautiful (2021) How African textiles and fashion have influenced the fashion industry | HelloBeautiful. [ONLINE], Available at:https://hellobeautiful.com/2939174/african-fashion-and-textiles-trend. Accessed 14 Oct 2021
9. UFO Themes (2021). African fashion shows create global impact | Empire Textiles BlogEmpire Textiles Blog. [ONLINE], available at: https://www.empiretextiles.com/blog/portfolio/african-fashion-shows-create-global-impact/. Accessed 15 Oct 2021
10. UFO Themes (2021) Stories and meanings woven into African Wax Prints | Empire Textiles BlogEmpire Textiles Blog. [ONLINE], available at: https://www.empiretextiles.com/blog/portfolio/stories-meanings-woven-african-wax-prints/. Accessed 14 Oct 2021
11. Africa Renewal (2021) African history told through fashionable printed clothing | Africa Renewal. [ONLINE], available at: https://www.un.org/africarenewal/magazine/september-2020/fashion-and-fabric-african-history-told-through-printed-wax-clothing. Accessed 15 Oct 2021.
12. Perez D (2019) How fast fashion invaded the Philippines retail market. [Online] Eco Warrior Princess. Available at: https://ecowarriorprincess.net/2019/03/how-fast-fashion-invaded-philippines-retail-market/. Accessed 15 Oct 2021
13. Esquiremag.ph (2021) Philippine Indigenous fabrics and its importance today. [ONLINE], available at: https://www.esquiremag.ph/culture/design/philippine-indigenous-fabrics-are-making-a-comeback-a00225-20171017-lfrm. Accessed 15 Oct 2021
14. Pineapple Industries (n.d.) PIÑA SILK. [ONLINE], available at: https://pineappleind.com/collections/pina-silk. Accessed 15 Oct 2021
15. Narra Studio (n.d.) The Inabel of Ilocos: Woven cloth for Everyday. [ONLINE], available at: https://narrastudio.com/blogs/journal/the-inabel-of-ilocos-woven-cloth-for-everyday
16. The Business of Fashion. 2022. How Designers Build Big Businesses Out of Old Fabric | BoF. [ONLINE], available at: https://www.businessoffashion.com/articles/sustainability/how-designers-build-big-businesses-out-of-old-fabric-bode-reformation-raeburn/. Accessed 15 Oct 2022
17. Manila Bulletin (2021). Championing local textiles the right way means a lot to our indigenous communities—Manila Bulletin. [ONLINE], available at: https://mb.com.ph/2021/01/25/championing-local-textiles-the-right-way-means-a-lot-to-our-indigenous-communities/. Accessed 15 Oct 2021

As African Textile and Fashion Grow and Go Global, How Can We Make Sure It Remain Sustainable?

Marc-Arthur Gaulithy, Christine Nantchouang Ngomedje, Priscila Bieni, and Ivan Coste-Manière

Abstract Favored by various factors such as the influence of the millennials, social media, the entertainment industry, the traditional African fabrics have gained a lot in influence in recent years in the fashion industry. These traditional textiles, are distinguished by the handmade technique, and the use of natural dyeing. They are manufactured by small communities of artisans with centennial know-how. In the face of globalization and pressure from the transition to a larger scale production, several questions are raised: is it possible to maintain these sustainable techniques as they are? Can the circular economy values found in the producing communities be sustained? Through our exploratory research, we have proceeded to a systematic review of the literature, generated a theoretical framework through which we analyzed the actions of local and international actors in the fashion industry. From this, emerged recommendations such as promotion of fair trade practices, protection and empowerment of artisans communities, and other hypotheses that will need to be tested through further research.

Keywords African fabrics · Fashion · Sustainability · Appropriation · Circular economy · Handmade · Natural dyeing

1 Introduction

When we speak of African textile, an emphasis is put on the term "traditional" to highlight the difference between what is known as "African wax", industrially made, and handmade, traditional African textile. This article will focus on the latter.

African wax prints originated in the nineteenth century during the colonial era, when European merchants familiar with African culture and their cloth-wrapping practice, detected the opportunity for a profitable business venture; the production of colored fabric patterns which would fit the African taste [17].

M.-A. Gaulithy · C. N. Ngomedje · P. Bieni · I. Coste-Manière (✉)
Skema Business School, Université Côte d'Azur, 60 Rue Dostoievski, 06902 Sophia Antipolis Cedex, France
e-mail: ivan.costemaniere@skema.edu

© The Author(s), under exclusive license to Springer Nature Singapore Pte Ltd. 2022
S. S. Muthu (ed.), *Sustainable Approaches in Textiles and Fashion*,
Sustainable Textiles: Production, Processing, Manufacturing & Chemistry,
https://doi.org/10.1007/978-981-19-0874-3_12

Although today the African wax prints remain largely consumed by Africans both locally and in the diaspora, they do not represent the indigenous identity of the African people. African textile and fashion have been around in the native continent for longer than we can imagine, the complexity of the continent itself—geographically, culturally, and historically—is mirrored in the substantial variety of textiles throughout the regions of a single country alone.

Moreover, Africans have contributed to the enrichment of fashion worldwide. For example, during slavery, African women took along to the Americas their cultural habits such as head wrapping; a habit that has extended to present times as a cultural affirmation. Another example is later in the 90s, when American volunteers from Peace Corps in Nigeria adopted the Yoruba luxury clothing known as Dashiki and continued wearing it once back home. Handmade West African textiles, such as Bogolanfini, Adinkra, and Kente have likewise traveled across the Atlantic and have become a trend [14].

The question now arising is, with the globalization of traditional African textile and the need for a larger production scale, how can we ensure that it will remain sustainable?

According to sustainability researchers Canan Saricam and Nazan Okur, sustainable fashion is clothing designed for a longer lifetime use, and produced in ethical production systems that use materials and processes not harmful to its environment [19].

Ever since their inception, traditional African textiles have been handwoven and their dye has been natural. Small community weavers and women dyers are behind its production, with however the knowledge of the latter slowly disappearing due to the pressure stemming from wax print textiles. Will this know-how stay around longer with the pressure of globalization? How will the fashion industry and global high fashion utilize these textiles while committing to the conservation of these traditional weaving, sewing and dyeing techniques?

2 Methodology

A combination of primary data and secondary data were used to gather information for this article. Primary data was gathered through interviewing people of interest, whereas the secondary data was collected by research of books, peer-reviewed scholarly articles, and reports. To find the data needed, exploratory research methods were used as we aimed to discuss an under researched topic. The data found was in majority qualitative.

2.1   Traditional African Textiles

There are almost as many traditional African textiles as there are tribes. But what the textiles all have in common is that they are usually handmade, of cotton, in some cases wild silk, dyed with natural pigments, and carrying with them a rich imagination like the history of the continent. Their wearing is often linked to particular events in the social life of the people from which they originate, such as tribal events, weddings, funerals, baptismal ceremonies, burials, and more.

For the purposes of this work we will highlight three of the most famous traditional African textiles: Bogolanfini, Adire and Kente.

2.2   Bogolanfini (Mud Cloth)

Bogolan is a hand-woven cotton textile, dyed with mud and other natural dyes, on which special patterns are drawn. Its best-known patterns are austere, geometric, usually in black and white. Originally, the Bogolanfini was intended to be worn as a shirt or tunic by hunters. Today it can be found in the form of a scarf, in decorative fabrics, or on the hips of a Beyoncé dressed in an African Amazon.

It is this technique of dyeing with mud that gave it its original name: Bogolanfini, which literally means in Bamana, "Fabric made with mud" [26]. Bamana live to the east and north of Bamako, and the best Bogolanfini work is done in Beledougou area [23].

This textile, generally made by communities of women, has certain particularities. Although it is now, one of the most famous ethnic fabric in the world [18], and worn in various forms by both men and women, the Bogolanfini is basically a *"deeply traditional fabric used by women at crucial stages of their lives such as excision, marriage and childbirth, and by hunters who believe that designs and patterns are imbued with a vital force, a powerful energy capable of protecting them in the bush"* [6].

Very little known in Western academic literature before the 1970s, the Bogolanfini owes its spotlight to Imperato and Shamir, who published a richly illustrated article in African Arts entitled "Bogolanfini: Mud Cloth of the Bamana of Mali", in which he provided the first comprehensive description of how this textile is created and the names and meanings of some of its most common designs [15].

Following him, several years later, a few researchers such as Picton & Mack, Victoria L. Rovine, Polakoff or Luke-Bone have enriched this literature through major publications that are references until today.

Alongside the scientific interest in Bogolanfini, Chris Seydou, a fashion designer of Malian origin, is the one who made Bogolanfini a famous textile on T's all over the world. Indeed, in the 80s and 90s, he presented several collections in several European capitals, notably in Paris where he worked and in New York. He changed the Bogolanfini from traditional ritual and hunter attire to diversified and modern garments such as motorcycle jackets, miniskirts, etc. [26].

As mentioned above, Imperato and Shamir described the most comprehensive Bogolanfini textile creation process. The production of Bogolanfini is purely manual process, and the whole process can take several weeks to complete. Locally produced cotton is combed and spun into a double-heddle loom into a narrow strip of about 15 cm in width. The strip is cut into shorter pieces. These strips are then joined selvedge with a whipstitch. The cloth is washed and dried in the sun. While the cloth is dried, a dye bath is concocted with leaves and branches of two different trees. These plants are pounded and soaked in water for 24 h, or boiled in water for few minutes. The cloth is then soaked in this solution and spread out to dry again in the sun. After that, the painting is done with mud that has been collected from ponds and left to ferment for a year. As the cloth is left to dry, the dark black turns grey. The cloth is then washed to remove the mud. The process of soaking in the leaf tea, painting, washing and drying is repeated as many times as necessary to obtain the colors and patterns needed. The areas that are not painted with mud, "the yellow base", are then discharged by applying soap or nowadays bleach, which makes it white again [33].

In recent times, Bogolanfini has found many other applications in response to the growth of tourism. Bogolanfini patterns are now also used in other commercial products such as coffee cups, curtains, towels, sheets, book covers, and wrapping paper [1] (Fig. 1).

2.3   *Adire Cloth*

Adire can be decomposed as following: $\bar{a}$ (to take) + di (to tie) + re (to dye). It is the indigo resist-dyed cloth renowned for its imaginative depictions of plants and people, birds and animals, is associated with Yoruba craftswomen in south-western Nigeria. The Yoruba word, adire, is defined in G. P. Bargery's AHausa-English Dictionary, as "*A Yoruba-made black-and-white cloth*" [24].

Adire fabric is easily recognizable by its indigo blue dye, on which various patterns made of mythical animals and birds, but also geometric patterns, are drawn [3]. Adire cloth is essentially handmade, thanks to a well-honed century-old technique. It is done in several stages and lasts on average one week, each stage being the work of a body of craftsmen qualified to carry out specific tasks. Basically, this confection is the work of women. They take care of the dyeing, spinning, and painting stages by hand, using accessories such as feathers. The men are instead involved in the planking and technical decorating stage [27].

Indeed, meticulously, using feathers, the craftswomen draw abstract or figurative patterns on a fabric. The batik, made from organic materials, is then dyed indigo blue. The color is obtained thanks to a mixture of leaf and cocoa pod ash. This makes it possible to obtain a shade of indigo blue.

Fig. 1 Bogolanfini, from Nakunte Diarra (Mali)
(Photo by Milbourne Karen, Smithsonian National Museum of African Art, 2013)

This highly prized textile was originally associated with the Yoruba craftswomen of south-western Nigeria. It was only in the 1930s, when an economic recession was taking its toll following the Great Depression, that this fabric was worn as a substitute for wax-printed fabrics which had more shimmering patterns and brighter colors. The Hausa (people of Northern Nigeria) thus appropriated the Yoruba Adire and reinterpreted it in their own way by adding new and more attractive patterns. Once the financial situation of the country improved in the 70s, the Hausa women went back to the practice of Wax prints, thus abandoning the Adire. It is the demand of tourists that will revive the Adire and evolve its design toward new styles. This tourist appeal will sustain its continued production to this day in the Kano area. Since then, the Adire has acquired great socio-cultural importance in Northern Nigeria [24].

The Adire cloth confection process consists in dyeing the basic fabrics which is generally Batik. Batik is immersed in cold water to soften it. Wax is then prepared. After melting, a wooden or foam stamp, designed in the desired pattern, is then dipped into the initially melted wax and stamped onto the batik in a horizontal or vertical form, which brings out the desired pattern. This is done as quickly as possible, because the wax cools and dries quickly on the application device [34]. Thereafter, the wax stamped batik is left to dry. After drying, the dye is prepared by mixing hot water and caustic soda. The mixture is stirred until it bubbles. Then the dye is

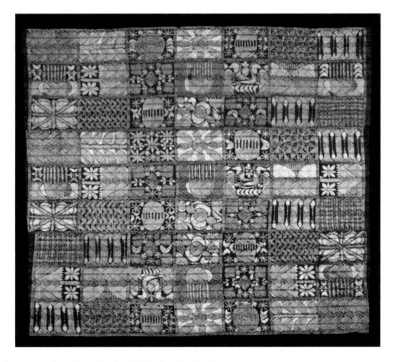

Fig. 2 Fine Adire Eleko Cloth with Ibadandun Design
(*Courtesy* Duncan Clarke, Adire African Textiles, 2016)

carefully added. Dyes were initially extracted from plants and leaves which were soaked for several days in order to extract the dye.

The waxed fabric is immersed in the mixture, making sure that all parts of the fabric are dyed. The excess dye is rinsed off and the fabric dried. The dyed fabric is then soaked for about thirty minutes, then rinsed and dried again. It will then be taken to planking, which consists of a kind of manual ironing done by beating the textile for hours with wood [27] (Fig. 2).

2.4 Kente Cloth

One of the most symbolic of these textiles of African origin is the Kente. Originally from Ghana, the Kente is also produced in Côte d'Ivoire. There are essentially two types according to the work of Boateng [8]: the Asante Kente and the Ewe Kente, Asante and Ewe being two tribes living in present-day Ghana and Côte d'Ivoire.

According to this work, Asante Kente is made up of single woven strips with alternating double woven panels so that when the strips are sewn together, the effect is like a checkerboard. It is distinguished by its bright colors and abstract patterns woven into the strips, unlike Ewe Kente in which the colors are more subdued and the patterns more realistic (see Fig. 1). The patterns used in the weaving of Asante Kente cloth have specific names; however, the cloth is usually named after the colors and the background pattern, which is often striped. Kente may also be named after historical figures and events as well as Asante values. For example, the design "*kyeretwie*", or leopard catcher, symbolizes courage, while "*aberewa ben*", or "wise old woman", indicates the respect given to older women in Asante society. While the designs of Kente cloth are generally abstract, weavers have expanded their aesthetic frameworks by weaving elements such as words, numbers and Adinkra symbols into cloth. These are usually strips intended for use as stoles or as decoration.

We see that Kente is not only a textile that is valued for its high aesthetic quality, but also because it is a medium to convey values and messages. It is also important to note the close link that the artisanal production of certain Kente designs had with the Asante royalty. Indeed, some craftsmen were put aside to produce Kente only for the use of the Asantehene (King of the Asante). This also makes it a conveyor of prestige values. Although today, this exclusivity tends to no longer exist, these unique designs of Kente, which legend has it that a woman created, remain some of the most expensive textiles in Africa, and highly symbolic of wealth, nobility, high social status and cultural sophistication, which are given as gifts at important ceremonies such as weddings. These fabrics are then kept as an inheritance that is passed on from mother to daughter or from fathers to sons [8, 13] (Fig. 3).

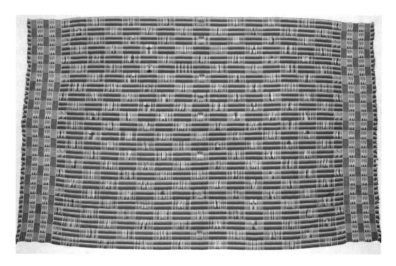

Fig. 3 Asante kente Cloth, 1935, Ghana, Cotton (British Museum) (Photo by Spring & Hudson, 2002)

2.5 A Transition to Be Supported

Today, these West African handmade textiles have crossed the Atlantic and Europe and have become "wearable chic" [14]. They have inspired several collections by famous houses and designers such as Jean-Paul Gaultier, Louis Vuitton, Yves Saint Laurent, John Galliano, Junya Watanabe [25], and TSE.

However, for some authors such as [5], the African fashion industry is internationally undervalued.

The African continent has long been considered as a supplier of sustainable and quality raw materials, such as cotton, or as a territory for the relocation of manufacturing plants in countries such as Mauritius or Ethiopia. It has also provided couturiers and designers of great talent, who have had an impact and will continue to have an impact on the global fashion and luxury industry with their particular know-how, as we will see later in this publication.

Vehicle of a rich imagination and a lifestyle that straddles tradition and modernity, the creations of African designers are the reflection of a society in perpetual change. They have the particularity of telling stories of cultural mixing and of sustainability philosophy that is found in the creation process of many of the fabrics and handmade materials with unique designs from craftsmen communities [7]. All this add a certain value to the global fashion industry.

Don't we see here some of the characteristics of the production of luxury products? Centuries-old tradition, craftsmanship, exclusivity, modernity, unique finishes?

However, this unique know-how in the world is threatened, as well by its success, as by the galloping urbanization of the African societies, that by globalization. Not being subject to mass production, traditional textile craftsmanship is suffering the full force of the dumping on the market of wax prints from Asia, but also prints mimicking some traditional textiles. The example of the "Kente-Like" industrially produced in China does not escape the informed observer [8].

Understandably, these attempts to mass-produce traditional African fabrics are interpreted, perhaps wrongly, as a subversion of culture, and are opposed by purists because of issues of cultural ownership, tradition or even ethics. The thorny issue of cultural appropriation or borrowing raised by Scafidi's work has resurfaced. According to him, cultural appropriation is defined as "*taking from a culture - not our own - intellectual property, cultural expression or artifacts, history or know-how*" [28]. Like Brooks or Buchlol, he believes that to do otherwise, then to retain the originality of a cultural artefact that one would like to borrow, could raise challenges and accusations of spoliation and other types of conflicts that often pit powerful and hegemonic majorities against disempowered minorities [10]. This is referred to by Brooks in "*Why Borrowing should Never Tip into Exploitation*" [9].

This poses a dilemma that needs to be resolved to enable the world of luxury and fashion to benefit from the richness of African textile culture. Another issue can be raised here: art and design can no longer ignore its social responsibilities. According to Odiboh [21] this issue must be considered as essential.

3 The Growth of Traditional African Textiles

We are therefore faced with a paradox. Traditional African textiles, whose symbolism and unique aesthetic richness are in increasing demand; opposite, is a fashion industry in search of a new economic model that is more sustainable and ethical. This increase in demand is influenced by various determinants, some of which we will analyze.

3.1 Drivers

3.1.1 The Genesis of African-Inspired Haute Couture

Certain luxury brands have recently been scrutinized for cultural appropriation faux-pas. Yet, this phenomenon is far from new. African textiles have served as inspiration to many European designers [1].

Over the past century, Europeans have shown interest in African cultures in various disciplines such as art and fashion. Their desire for creativity and originality led to stylistic borrowing by multiple high-end design houses. As Ethiopian model Anna Getaneh mentioned, Africa has a lot to offer and these mainstream designers are much aware of this [29]. The uniqueness of African textiles has inspired creations such as Madame Carven's African collection in the 1950s, Yves Saint Laurent's "Bambara" dresses or Kenzo's bogolan-inspired jackets and shirts. 1997 was dubbed "The Year of Africa" in Europe's fashion industry due to the numerous stylistic references to Africa [17].

3.1.2 The Role of the Entertainment Industry

Another major driver of the growth observed in the industry is initiated by African-Americans. The Black Lives Matter movement, along with the need for African-Americans to find and go back to their roots have played a consequent role in the shift of mentality that occurred. In today's context where African and African-Americans are proud of their origin, it is important to note that being African or having African roots was often a subject of mockery. The Black community went from almost denying their African roots to fully embracing their African identity. It is not only a trend effect anymore, it is part of a strong political move.

This movement was mostly triggered by influential American celebrities. Their actions have significantly encouraged their peers to discover their African identity with DNA testing. As a result, celebrities such as Erykah Badu and Oprah Winfrey finding out their Cameroonian origins, or others even taking the citizenship of their country of origin such as Samuel L. Jackson. The 2019 Year of the Return has also supported this movement. Ghana successfully hosted this year-long program to commemorate the 400th anniversary of the arrival of the first recorded African slaves in the United States. During this event, Ghana's President, His Excellency Akufo Addo granted citizenship to members of the diaspora, further confirming that the country (and the continent) is their home [31]. Identity and apparel being intrinsically linked, this shift had a considerable impact on the perception of African textiles. Recognition was boosted even further when influential people—especially from the entertainment industry—started wearing African textiles or designs made by Africans. When Beyoncé Knowles wore a coat designed by South Africa-based design house Kisua in 2014, it was immediately sold out [11]. Former First Lady of the United States of America, Michelle Obama and actress Lupita Nyong'o have both worn designs by Lagos-based Maki Oh. Cameroonian designer Claude Kameni is behind Viola Davis' 2021 Golden Globes Award dress, and one of her designs was featured in Coming 2 America. Moreover, the premiere of the Marvel Comics-based movie Black Panther in 2018 was a groundbreaking display of various African fabrics and attires, with actors wearing dashikis, boubous and Daniel Kaluuya wearing a kanzu, a traditional attire of Uganda [30]. Having these influential household names embracing their African roots with pride prompted a cumulative effect that has changed stereotypes and is now triggering a permanent change. As Samuel Mensah, founder of Kisua stated, African fashion being coveted by Americans shows a transition to the industry becoming mainstream [11].

3.1.3   A Change in Customer Behavior: The Millennials Case

Traditional African textiles are known for being made in a sustainable way. As consumers are becoming more aware of the impact of their purchase decisions, it is important to mention that Millennials are indirectly playing a huge role in the global shift to sustainability.

Millennials have been the subject of multiple researches about their characteristics, their behavior and how businesses can meet their needs. This generation, being the largest demographic and the largest consumer group in Africa and in the world today, their impact on consumer trends cannot be neglected. In addition to being tech-savvy, they are the most sustainability-conscious generation to date. They are socially conscious and are loyal to companies that are committed to doing their part when it comes to corporate social responsibility. Their consumption trends and tastes are anchored in the Sustainable Development Goals (SDGs), including climate change, equality, and poverty. A 2018 study by Nielsen [20] showed that 53% would be willing to give up on a brand to instead purchase a more environmentally friendly alternative. They also prefer to pay more for environmentally friendly products. Overall, there is an emphasis on supporting companies with clear and honest sustainable practices. When it comes to luxury, Bain & Company predicts that Millennials and Gen Z consumers will account for half of the total global luxury sales by 2025 [12]. As this segment continues to grow, luxury brands focus on them because of the long-term potential they represent. Yet, they desire that luxury brands align with their sustainability-driven values. When making a luxury purchase decision, they seek brands that will provide a way for self-expression, and they spend more on socially responsible and environmentally conscious brands.

3.2 Some African Fashion Designer Focused on Sustainability

The growth of the African traditional textile industry has been spurred by designers that desired to transform tradition into a global effect. To do so, two approaches can be mentioned.

The first approach is to divert these clothes from their traditional use. Indigenous textiles have a meaning and are often used in precise cultural settings. The challenge here is to adapt to Western codes. Chris Seydou's work is a notable illustration of this. The creator's designs with indigenous textiles such as the Malian bogolan-fini illustrates that local tradition and global markets can be bridged. The designer reimagined the bologan cloth by creating his own versions of the material out of respect for the cloth's ritual significance in rural Mali. These versions were turned into a large apparel of clothes such as skirts, coats, or even fitted suits as the one worn by Former First Lady Adame Ba Konaré for the opening of a film festival in Marseille in 1993 [26]. He later collaborated with the Malian "Industrie Textile" by designing a bogolan-inspired fabric that was sold between 1990 and 1991. Fashion designer Titi Ademola had the same view about promoting Ghanaian fabrics to the world, not only Africa. With creativity and sustainability as the driving forces behind her Accra-based KIKI Clothing brand, her collection is inspired by everyday life and caters to different people with different lifestyles.

African textiles are at the heart of an ongoing evolution. Contemporary African designers prove that it is possible to transform traditional textiles into luxury, while incorporating notions such as recycling and digitalization. African designers are now making sustainability the center of their creativity. A notable example is Parisian fashion house XULY Bët, founded by Malian designer Lamine Kouyaté in 1991. This pioneer of sustainability is best known for transforming recycled clothing by reshaping them into high fashion pieces. The designer purposely leaves the original labels found on the clothes, providing evidence of the clothes' previous life. The designer works with dyers in Mali to create fabrics for his other clothing lines, which fits into his sustainable approach [25].

Loza Maléombho is another designer worth mentioning. The self-titled brand created by the Ivorian American designer is a cooperation between West African royalty and New York City's urban style. Her creations are also a paradox, as they are a bridge between a futuristic approach and a traditional approach on the other hand. As stated by herself in an interview, her intention was to create clothes that were a mix of contemporary and traditional elements that women around the world could identify with. Sustainability is at the core of her brand and her approach to growth puts an emphasis on digital. Now based in Côte d'Ivoire, she works with artisans that incorporate their savoir-faire throughout her collections.

Given the Millennials' mindset about sustainability and their strong affiliation to technology, it is correct to conclude that this generation would be much more inclined to supporting the African traditional industry's growth. As seen previously, many design houses have their supply chain rooted in Africa. As the most ethnically diverse generation, Millennials worldwide are not only creating a change environmentally and ethically; they are also proving that it is possible to be fully anchored in globalization while fully embracing your roots. This has been the case for African and African American Millennials. We can also expect this segment to turn to small or medium-sized local businesses providing luxury goods made in an ethical way.

The industry is gaining momentum and further grow is already anticipated. Transforming indigenous African textiles into luxury goods implies an even greater challenge, which is to keep sustainability as a core value. We will now be studying the economic, social, and environmental implications of creating luxury African fashion at a greater scale.

4 Sustainability Implications of Growth

The current trends and changes in the textile industry raise a new issue of concern, which is to keep the industry sustainable. Despite the potential for this sector to grow in a sustainable way, major challenges are still faced. These challenges can be classified into categories that correspond to the three pillars of sustainability: economic, social, and environmental. To understand the implications of African indigenous textiles making their way into sustainable luxury, it is important to study each of those aspects.

4.1 Economy and Social

Even though today's fashion industry is valued at US$3 trillion and employs more than 300 million people globally, the processes used in this sector require urgent transformation. Resources used to produce textiles are finite and toxic chemicals are often used. On the other hand, there is an important issue of waste. Products are either unsold, unused, or discarded despite still being in good condition. The rise of fast fashion tends to heighten the issue, with large volumes of low-quality items being produced but not easily recyclable. To tackle all these problems, a brand-new concept has been defined: Circular Fashion. The term "Circular Fashion" was shown in 2014, and has rapidly become one of fashion's most embraced sustainability concepts since late 2018. More recently the *"Circular Fashion Report"*, *the* result of the collective work of industrialists (Lablaco, Startupbootcamp FashionTech), academics (ESSEC

Business School, Wageningen University), and consulting firms (PWC, Anthesis, Rödl & Partner), estimates the market for the circular fashion at US$5 trillion [16].

According to the African Development Bank (ADB), 90% of fashion businesses in Africa are small to medium-sized but lack resources, necessary education and means of industrialization. Shipping fabric and raw materials induce high transport costs for these businesses. As these small institutions cannot compete against worldwide-renowned design houses, they are often faced with loss of intellectual property due to the high cost of copyright. Moreover, these businesses are not supported by the government [2].

These issues, coupled with others such as limited access to information, including trends, Regulatory issues(tariffs), and exchange rates affect exports …, illustrate the need for a transition to a circular economy for textiles.

The production of indigenous textiles is currently in the hands of small rural communities in Africa. These communities are usually composed of women or artisans with years of experience in their craft. The methods of production are well anchored in sustainability. But due to the high pressure of wax-printed textiles and the counterfeits, these small rural businesses are in danger. Few studies are able to estimate the real market size of the indigenous textiles in Africa. A study by Michelle Okyere, demonstrates that Kente imitations have reduced the profit being obtained by Ghanaian Kente producers. According to this study, the annual contribution of Kente in the textile industry to Ghana's GDP declined from USD 179,50 million in 1994 to USD 53.5 million in 2011 [22]. The same study mentions that, in order to protect the Ghanaian Kente, the concept of Geographical Indicators (GI) for non-food items have been introduced by Okyere and Dr. Denoncourt, at the World Intellectual Property Organization (WIPO).

4.2 Environment

The contribution of the African countries in the textile and fashion industry can be divided into two main parts: The producers of raw material, notably cotton, which are the countries of West Africa (Benin, Ivory Coast, Mali, Burkina Faso) on one hand; and on the other hand, those that can be called factory countries such as Ethiopia, Mauritius, South Africa… Depending on which of the above categories one falls into, the environmental challenges are different. The cotton producing countries for example face issues of use of pesticides and other products of treatment of the raw material such as fertilizers, the use of GMOs… Countries like Côte d'Ivoire, Cameroon, Ethiopia, Ghana, Mozambique, or Tanzania belong to the Cotton made in Africa (CmiA) initiative. Cotton made in Africa (CmiA) is an initiative of the Aid by Trade Foundation (AbTF) which aims to promote economically, environmentally, and production of socially sustainable cotton in Africa. Mali, Benin, and Senegal are members of the *Fairtrade* Initiative [32].

For factory countries such as Mauritius, South Africa, or Ethiopia, the major challenge is the cost of labor and recycling. The Ellen MacArthur Foundation evaluates that less than 1% of all clothing is recycled into garments worldwide. At the same time, approximately 12.5% of the global fashion market has publicly committed to circularity by signing the *Circular Fashion System Commitment*. Recycling cotton recycling is a path toward a more circular fashion industry [32].

The African Development Bank has launched an initiative in 2019 called *"Fashionomics"*. This initiative is articulated around the *"High-5 Agenda"*, one of which, *"Power Africa"*, addresses the issue of sustainable energy in the textile and fashion industries. Companies are encouraged to use renewable energy and green jobs, which is a big step toward sustainable development. Examples include recycling of textile products, minimization of toxic substances, alternatives to existing raw materials, waste reduction, reducing of energy consumption, renewable energy and consideration of the life cycle of products.

4.3 Recommendations

In January 2021, under the leadership of Ghanaian-born American designer Virgil Abloh, Louis Vuitton paraded the Fall-Winter 2021 men's collection in drapes that strongly resemble Kente. To add insult to injury, on the May 2021 cover of Vogue, Amanda Gorman posed in an outfit made from the same drape. What looked like a wonderful "publicity stunt" for the prestigious Ghanaian textile turned into a loud scandal. Some critics, such as Pierre Antoine Vettorello, believe that this was a show of hegemonic appropriation on the part of the prestigious luxury house. According to him, Louis Vuitton missed a great opportunity to pay tribute to Ghana and its rich culture.

By refusing to have its Kente cloths produced by Ghanaian craftsmen, and by producing them in its workshop, with the addition of its logo branding, it was illegally appropriating a heritage that is not its own, and which belongs to a whole community. For a company that constantly emphasizes its use of the most qualified craftsmen, this affair did not go down well [35].

This situation, which is far from an isolated event in the fashion industry, highlights one of the strategic challenges expressed by this work: African traditional textile artisan communities must be protected; they and their know-how.

For traditional African textiles to continue to bring that touch of sophistication, that *"je ne sais quoi"*, *to* the global fashion industry, the producers and their methods of production must be protected and recognized as such. Ghana, for example, has applied to the WIPO for a Protected Geographical Indication (GI) for Kente. The sourcing of its textiles should be done in fair trade. These artisans should also see their production capacities strengthened, both with the help of the government and the industries that solicit them, through the CSR.

Top management of major brands must go beyond mere expressions of good intentions and comply with the various commitments made during the numerous initiatives to promote sustainability in the luxury and fashion industry.

5 Conclusion

The Parisian fashion fortnight ended this year 2021 with a fashion show of the Cameroonian stylist Imane Ayissi near the Champs Elysée, thus closing the week of haute couture. In January 2020, he became the first designer from sub-Saharan Africa to join the elite club of Paris haute couture. He had introduced the public to little-known African know-how: *tie and dye* dyed in Cameroon; Kente, traditional weavings of the Akan ethnic group found in Ghana and Côte d'Ivoire and originally worn by the nobility; or Obom, a vegetable skin produced from tree bark, which he used to adorn evening wear. During this coronation, he explained to Agence France Presse his desire to open "*a new way for* Africa" and find an "*alternative way to make luxury fashion*". He also expressed his rejection of wax prints, which according to him, are not a reflection of his African identity.

> When we talk about African fashion, it's always wax, which is a real shame…because it's killing our own heritage. [4]

We could not find a better conclusion for this article, whose objective is to show that, if all the actors of the value chain play their part, the textile of African origin can very well become globalized, and enrich the luxury and fashion industry, while remaining sustainable.

References

1. Acquaye R (2018) Exploring Indigenous West African fabric design in the context of contemporary global commercial production. University of Southampton, s.l
2. ADB (2020) Investing in creative industries: fashionomics. [En ligne]. Available at: https://www.afdb.org/fileadmin/uploads/afdb/Documents/Generic-Documents/Fashionom ics_creative_industries_executive_summary_brochure.pdf. Accessed 16 Oct 2021
3. Aero MO, Kalilu RO (2013) Origin of and visual semiotics in Yoruba textile of Adire. Art and Design Stud 12:2–224
4. AFP (2020). Paris (AFP). Imane Ayissi, couturier du patrimoine africain, sauf le wax [En ligne]. Available at: https://www.lecourriercauchois.fr/actualite-225207-paris-afp-imane-ayi ssi-couturier-du-patrimoine-africain-sauf-le-wax. Accès le 16 Oct 2021
5. Aziz M, Alexandre-Leclair L., Salloum C (2019) The fashion industry in Africa: a global vision of the sector. Dans: Moreno-Gavara C, Jimenez-Zarco AI (éds) Sustainable fashion empowering African women entrepreneurs in fashion industry. s.l.:s.n
6. Barton W (2007) African Mud Cloth. The Bogolanfini art tradition of Gneli Traoré of Mali (review). Afr Stud Rev 50:210–211. https://doi.org/10.1353/arw.2005.0091
7. Bell S, Morse S (2013) Measuring sustainability: learning from doing. Routledge, New York
8. Boateng B (2011) The copyright thing doesn't work here. Adinkra and Kente Cloth and Intellectual Property in Ghana. University of Minnesota Press, Minneapolis
9. Brooks X (2016) Cultural appropriation. Mix Mag 45:17.
10. Buchloh B (2009) Parody and Appropriation in Francis Picabia, Pop and Sigmar Polke. Dans: Appropriation. Whitechapel Gallery Ltd., London, p 178
11. Chutel L, Lijadu K (2018) What happens to African designers when Beyonce and other stars wear your clothes [Online]. Available at: https://qz.com/africa/1382554/beyonce-theresa-may-michelle-obama-get-into-african-fashion/. Accessed 14 Oct 2021

12. Claudio D, Nieto DV, Davis-Peccoud J, Capellini M (2021) *LuxCo* 2030: a vision of sustainable luxury. Bain & Company Inc., s.l
13. Contemporary African Art (2016) Kente Cloth [En ligne]. Available at: https://www.contem porary-african-art.com/kente-cloth.html. Accessed 15 Oct 2021
14. Gott S, Loughran K (2010) Contemporary African fashion. Indiana University Press, s.l
15. Imperato PJ, Shamir M (1970) Bokolanfini: Mudcloth of the Bamana of Mali. Afr Arts 3(4):32–42
16. Lablaco (2020). Year Zero. Circular Fashion Report 2020 [En ligne]. Available at: https://doc send.com/view/63avn4jc3ztb952w. Access 14 Oct 2021
17. Loughran K (2009) The idea of Africa in European high fashion: global dialogues. Fash Theory 13(12):243–272
18. Luke-Boone R (2001) African fabrics. Krause, Lola
19. Muthu SS (2019) Fast fashion, fashion brands and sustainable consumption. Springer Singapore, Singapore
20. Nielsen (2018) Was 2018 the year of the influential sustainable consumer? [En ligne]. Available at: https://nielseniq.com/global/en/insights/analysis/2018/was-2018-the-year-of-the-inf luential-sustainable-consumer/. Accessed in October 2021
21. Odiboh F (2005). Is still largely orchestrated by people Leopold Senghor (1967). Nigerian J of Art 4(1–2):45
22. Okyere M, Denoncourt J (2021) Protecting Ghana's intellectual property rights in Kente textiles: the case for geographical indications. J Intellect Prop Law Pract 16(4–5):415–425
23. Polakoff C (1982) African textiles and dyeing techniques. Routledge & Kegan Paul, London
24. Renne EP (2020) Reinterpreting Adire cloth in Northern Nigeria. Textile Hist, 1–26
25. Rovine VL (2015) African fashion global style: histories, innovations, and ideas you can wear. Indiana University Press, Bloomington and Indianapolis
26. Rovine V (1997) Bogolanfini in Bamako. Afr Arts 30(1):40–55
27. Saheed ZS (2013) Adire textile: a cultural heritage and entrepreneurial craft in Egbaland, Nigeria. Int J Small Bus Entrep Res 1(1):11–18
28. Scafidi S (2005) Who owns culture? Appropriation and authenticity in American law. Rutgers University Press, New Jersey
29. Shivers ND (2011) Fashion as performance: influencing future trends and building new audiences. Dans: Fashion forward. Brill, s.l, pp 405–417
30. Smith J (2018) The revolutionary power of Black Panther [Online]. Available at: https://time.com/black-panther/. Accessed 14 Oct 2021
31. Tetteh B (2020) Beyond the year of return: Africa and the diaspora must forge closer ties [Online]. Available at: https://www.un.org/africarenewal/magazine/september-2020/beyond-year-return-africa-and-diaspora-must-forge-closer-ties. Accessed on 14 Oct 2021
32. Textile Exchange (2020). 2025 sustainable cotton challenge. Second Annual report 2020, s.l.: TextileExchange.org
33. Toerien E (2003) Mud cloth from Mali: its making and use. J Family Ecol Consum Sci/Tydskrif vir Gesinsekologie en Verbruikerswetenskappe 31. https://doi.org/10.4314/jfecs.v31i1.52840.
34. Tomori S (2011) The impact of Adire on the cultural heritage and economic growth of Ogun state. Unpublished thesis of the Achievers University, Owo
35. Vettorello PA (2021) Louis Vuitton and African textiles: the case of Kente [En ligne]. Available at: https://www.pierreantoinev.com/kenteandlouisvuitton. Accès le 16 Oct 2021

Ethical Fashion: A Route to Social and Environmental Well-Being

Yamini Jhanji Dhir

Abstract Ethical Fashion pertains to adoption of sustainable approaches at all levels of fashion supply chain spanning designing, production, retailing and purchasing. A range of social and environmental issues like working conditions, exploitation and regularization of wages for workforce, fair trade, sustainable production involving watchful utilization of natural resources, animal welfare and minimizing environmental impacts constitute the umbrella of ethical fashion. However, the greatest threat to ethical practices is inception of fast fashion credited to the globalization and industrialized methods of growing natural fibres. Consequently, the high street clothing becomes affordable and easily available to masses owing to quick and cheap, bulk production in bulk quantities. Ethical Fashion targets the glitches and adversities associated with operational practises of fashion industry like workforce exploitation, deleterious impacts to environment, the utilization of hazardous chemicals, dyestuffs and waste, exploitation and wasteful usage of natural resources and inflicting cruelty on animals for leather, fur, ivory, etc. The adoption of ethical fashion in design and production phase can play a pivotal role in averting the deleterious social and environmental adversities owing to stringent control on the whole supply chain from designing stage, material procurement, production process, to delivery and life cycle analysis of end product. Furthermore, major apparel and accessory brands sensing the consumers' preferences and the worldwide clamour for sustainable and ethical fashion, have been adopting ethical approaches in order to adapt to changing scenario. The fashion brands in the pursuit to follow ethical principles practise pre-evaluation of suppliers for amenities, technical prowess, product quality, customer service, innovation and compliance of codes of conduct and contracts for rating the suppliers on above parameters. Moreover, the design process is positively impacted as brands improve procurement practices at their end, reducing their total costs of introducing new styles to market, a fast style inception. The practices, if followed at the design stage, can lead to fewer supplier compliance violations, thus offering overall benefits to the design process. Furthermore, designers can play a vital role in diverting textile waste away from landfill and prolonging the garment and accessory's life cycle by switching over to reusable textile waste, trims and notions, while designing some

Y. J. Dhir (✉)
Fashion & Apparel Engineering, TIT&S, Bhiwani, India

S. S. Muthu (ed.), *Sustainable Approaches in Textiles and Fashion*,
Sustainable Textiles: Production, Processing, Manufacturing & Chemistry,
https://doi.org/10.1007/978-981-19-0874-3_13

of their upcoming style lines. The scrap, textile waste and discarded clothes, trims and accessories, if used strategically and reformed into new apparel and accessory styles, not only add to profit margins for those involved in the supply chain but also enables the process to achieve sustainability. Accordingly, a gamut of fashion designers and brands namely Anita Dongre, Sandra Sandor, Katie Jones, Maggie Marilyn, Gautam Gupta, Paromita Bannerjee and Thief & Bandit, The Little Market, Patagonia, Remake score, Love Stories Bali, Nicora, Doodlage, Pero, Grass roots, Nicobar, respectively have been trailblazers in creating apparel and accessory line based on recycling, reuse and upcycling principles. Additionally, the newly introduced concepts of capsule fashion, convertible garments and gender fluid clothing are all supporting the ethical commitments of the designers and manufacturers. The chapter presents a detailed review on ethical fashion as a pathway for social and environmental well-being.

Keywords Brands · Designers · Ethical · Capsule · Gender-neutral · Convertible · Social · Workforce · Textile · Fashion · Sustainability · Supply chain · Environment

1 Ethical Fashion

Ethical fashion refers to fashion that aims at subverting the deleterious impact of garment production on flaura, fauna, masses and the planet. Ethical fashion is about being kind and empathetic to the planet and society during each garment manufacturing stage. The initiatives towards promotion of ethical fashion can pave the way for creation and strengthening of social enterprises in emerging economies thereby establishing meaningful connect between international fashion brands with well-versed indigenous designers, artisans and micro-producers.

2 Ethical vis a vis Fast Fashion

Ethical and fast fashion can be demarcated based on production processes and lifetime of the apparels and accessories The design phase of clothing produced ethically will encompass consideration of several factors besides functionality and serviceability such as ability to withstand trends, and versatility. The fast fashion model, contrastingly primarily focuses on selling the largest quantities as rapidly as possible following the concept of 52 mini seasons rather than the conventional four of spring, summer, autumn and winter. Target, H&M, Zara, Forever 21 and Primark are some of the apparel brands that follow the concept of fast fashion.

3 Adversities of Fast Fashion on Environment and Society

The concept of fast fashion or quick fashion is advantageous for all stakeholders namely apparel and accessory designers, manufacturers, suppliers and retailers owing to enhanced sales and profitability achieved through affordable design collections. The fast fashion model is undoubtedly hitting the main stream market where fashion companies and distribution networks largely focus on design upgradation of their collection along with low production costs, fast flow and reasonably priced merchandise.

Additionally, consumers are also enticed by the fast fashion merchandise since they are offered with an assortment of trendy, affordable and accessible ensembles.

Fast fashion portrays quite distinctive characteristics by offering massive style diversification unlike the conventional large-scale and low-cost apparel production process. The main drivers of fast fashion include responsiveness to changing trends, styles and consumers; preferences and demands. Clothing manufacturers and retailers in order to scale up their businesses and meet the expectations of target consumers, offer product diversification and launch exclusive, refreshing and affordable merchandise lines. However, in so doing, they become reckless and turn a blind eye towards the negative impact of fast paced and unsustainable designing, sourcing, procurement, production, packaging and shipment on environment and society.

Zara, a top-notch women western wear brand promotes fast fashion by following 20 micro seasons per year in contrast to traditional four seasons thereby offering an array of quirky yet economical merchandise for consumers to fill their closets. The synchronization of production-delivery phase leads to reduction in batch sizes coupled with shorter lead times thus overburdening the workforce. The pressure of meeting deadlines forces them to work beyond their means thus taking a toll on health and social well-being of blue collared staff.

It can thus be recapitulated those manufacturers and retailers in the pursuit to follow fast fashion concepts have been negligent to sustainable alternatives and the impact of micro seasons in terms of exploitation of natural resources, the usage of toxic, chemical products, poor labour conditions and the trend of outsourcing manufacturing and job work to low labour cost countries. It thus becomes imperative to transform the fast fashion model to ethical and sustainable one to relinquish the environmental and societal hazards inflicted by fast fashion.

The climatic change related challenges can be accounted to Fast fashion as it is economical, fad style thereby spiralling into ever shorter and faster production and sales cycles.

3.1 Problems Associated with Fast Fashion Include

- Deleterious environmental impacts in the form of soil, water and air pollution, global warming

- Massive consumption and exploitation of natural resources
- Complex supply chain.

Ethical fashion encompasses due consideration of social altruism, well-being and protection of workforce at each stage of garment supply chain. The human centred approach of ethical fashion assists in assessment of impact of each process of supply chain on workforce.

Ethical fashion unlike the fast fashion does not inflict any abuse or exploitation, sweatshops or child labour in the entire supply chain.

It wont be inappropriate to point out that the former deals with masses, their well-being and the planet while the latter mainly focuses on profitability.

The characteristics of ethical company can be broadly enlisted as:

- Transparency with the ecological procedures
- Fair wages and conducive work environment
- Traceability of end products
- Transparency about generated wastage

4 Need of the Hour—Ethical and Sustainable Procurement in Fashion Industry

The concept of ethical and sustainable procurement and clean clothing is emerging as a crucial concept in apparel production as modern-day consumers give precedence to environmental impact of attires they use along with functional, economic and aesthetic aspects. All the stake holders associated with fashion industry are thus bound to comprehend the hazardous impacts of unsustainable production processes and commit to adopt ethical practices in their supply chain. The focus, is thus on procurement of such textile raw materials that are nonhazardous, do not result in human health-related issues and in no way violate the social rights of employees working in these production units.

The alarming and menacing effect of fast fashion and unethical approaches on environment and society has been elaborately discussed in the previous section. Having discussed that, not just the wrongful selection of raw materials used for clothing production can deplete the natural, renewable energy sources but the excessive utilization of toxic and hazardous dyestuff and auxiliaries, pesticides, fertilizers, chemicals, during various production and processing stages have significant role in negatively impacting the environment by damage to flora and fauna, global warming, extreme climatic change soil infertility, air and water pollution.

The adoption of ethical principles in fashion industry is all the more challenging with garment recycling a major cause of concern. Reprocessing and recycling of textiles is problematic owing to abundant utilization of petro-chemical based synthetic fibres for garment production. The lack of SOP for disposal of garments after their life cycle results in discarded garments ending up as landfills thereby adding to plastic content in soil and water.

The adversities offered to environment during clothing production can thus be pacified by formulation and implementation of some strong strategies and in turn introduction of sustainable concepts at different levels of textile and fashion supply chain.

5 Importance of Ethical Fashion

The emergence of fast fashion with trends changing every now and then with minimal workforce and environmental consideration accounts for the deep rooted unsustainability in the entire garment supply chain. The need of the hour is thus saving the planet and human lives by revamping the supply chain by integration of ethical principles at each manufacturing stage.

6 Sustainable vis a vis Ethical Fashion

While ethical fashion primarily focuses on social altruism, well-being and protection of workforce working at different levels in industrial set up, the quintessence of sustainable fashion pertains to environmental aspect of manufacturing processes. Although the central idea of sustainable approach does not revolve as much around well-being of labour force however the impact of fashion on human health in an environmental context falls in the sustainable domain. Nevertheless, the underlying ideology of both ethical as well as sustainable fashion is to analyse effective measures of strengthening the garment production process by creation of eco-friendly mannerism and culture thereby attaining environmental justice.

Values and ethics are fundamental concepts of sustainable fashion. Ethics comprises of making deliberate sustainable choices in the designing, manufacturing processes, consideration of the business models and profit-making options. Value in sustainable fashion pertains to the impact of designing and industrial manufacturing processes on the environment, human health and social well-being. The exploitation of workers is on the rise owing to relaxed labour laws allowing the blue collared staff to work for long hours and that too at low wages and perilous working conditions. Garment workers particularly females fall prey to prejudicial working conditions and thus there is an ardent need to pay due consideration to working hours, wages, overtime and incentives for the workforce who are undoubtedly the back bone of the industry. Accordingly, several organizations like Sharehope are committed to work for incorporation of ethical principles in the supply chain and pledge to ameliorate labour standards for garment workers. Furthermore, apart from fostering fair working conditions in industry, the designers, manufacturers, brands and retailers should attempt to cultivate eco-friendly practices for holistic environmental and societal well-being.

Some effective measures for incorporation of ethical practises in textile and fashion supply chain encompasses

- judicious and minimal utilization of natural resources and energy sources in production
- Following the 4 R concept of reduction, reusage, recycling and repairing garments.
- Consumers should refrain from mindless shopping and need to be more watchful while making purchases.
- Styling by mixing and matching, following the capsule fashion and swapping fashion ensembles with peer group.
- Sustainable and ethical shopping by opting for second-hand merchandise primarily via virtual purchases on websites such as ThredUp and usage directories like The Thrift Shopper for in-person shopping.
- Awareness about sustainable brands and making conscious efforts to support sustainable brands that pass Remake's criteria.
- Concept of intersectional environmentalism for better understanding on correlation between privilege and sustainable fashion.
- Mindfulness about greenwashing.

Ethical fashion apart from ensuing fair wages, regularized working hours and safe working conditions also allow the consumers to perceive and define ethical fashion in accordance with their individual preferences and reliability on the brand.

The violations that result in unethical practises in textile and fashion supply chain involve:

- Unstandardized, irregular wages and forced overtime
- Child labour
- Exploitation of migrants & pregnant females
- Gender Discrimination
- Verbal, sexual and physical abuse
- Negligence of workforce safety measures & hazardous working environment.

The ideologies of sustainable and ethical fashion overlay as both follow holistic approach of maximizing the benefits to industry and society along with minimal environmental impacts by due consideration of each manufacturing stage right from sourcing, designing, manufacturing to packaging and retailing of apparels and accessories.

7   Apparel and Accessory Brands Designing Ethically

A gamut of fashion designers and brands namely Anita Dongre, Sandra Sandor, Katie Jones, Maggie Marilyn, Gautam Gupta, Paromita Bannerjee and Insecta, Darono, Flippa, Adidas, H&M, Zara, Mango, Indian Terrain, Brandili, Ecosimple, Patagonia, Doodlage, Pero, Grass roots, Nicobar, respectively have been trailblazers in creating apparel and accessory line based on recycling, reuse and upcycling principles.

Insecta, a Brazilian footwear brand offers an assortment of exclusive, vegan and ecological footwear for their elite and loyal consumers by utilization of vintage clothing and recycled plastic (PET) bottles. The company is proactive in following other sustainable principles such as tree plantation for neutralization of CO_2 emissions. Furthermore, they have changed the shipment and delivery modes by switching over to most eco-friendly alternative like bicycles.

Darono, a Portugal based brand experiments with environment-friendly polymeric fibres in their interior design collection.

Vintage for a cause support professional unemployed females by selling vintage clothing developed from old, discarded and dumped fashion ensembles.

Filippa K. encourages shoppers to purchase second-hand clothes which the brand offers in retail outlets thus playing a pivotal role in effective waste management. The consumers are further offered with long-term warranties and instructions regarding free repair, replacement and upcycling of clothing.

Some smaller brands following the league, have already adopted this technique and cater to customized orders of customers without indulging in overproduction. Accordingly, fashion brands are formulating revised policies by introducing the sustainable procurement concepts.

Furthermore, brands are hiring sustainable advisory firms to advise them regarding the best possible sustainable alternatives at optimum prices.

Adidas has been pioneer in sustainable procurement by recycling fibers from waste plastic and thereby using the recycled fibers for the design and development of sneaker sports shoes. Likewise, brands such as Zara, H&M and Mango are coming up with sustainable high street ranges by complying with the principles of sustainability and thus ensuring maximum utilization of environmentally friendly fabrics and processes.

Another apparel brand that follows ethical principles in design and development of their merchandise is H&M. The brand has introduced Conscious collection that primarily focuses on minimizing the environmental impact by switching over to sustainable raw materials like organic cotton or recycled polyester. The brand also encourages the consumers to recycle their unwanted clothing at H&M stores, alluring them with discounts on new merchandise in lieu of their old clothing.

Likewise, Lindex came up with its summer collection, Sustainable Choice, solely devoted to sustainable materials.

Mango's Committed Collection offers design collection comprising of certified organic and recycled cotton, polyester and tencel fabrics. The brand is further developing an internal tool to determine its water footprint in order to reduce water consumption. Small and medium-sized fashion brands have become equally cautious of their consumers' sustainable requirements and are, thus, committed to procure, manufacture and deliver products and services in accordance with principles dictating sustainable procurement.

A children's clothing brand, Brandili Textile, has adopted a sustainable stance by prudent utilization of natural resources. Ecological yarns are being produced by reusing textile waste, textile materials and polyethylene terephthalate (PET) bottles.

The company follows reverse logistics by formulation of post-consumer distribution channels.

Ecosimple, an American-based clothing brand, combines technology with sustainability by utilization of 100 percent-recycled PES from bottles, discarded clothing and fabric leftovers from garment factories and waste from spinning processes. Moreover, the brands are taking extra care in the selection of suppliers to ensure that sustainable procurement is followed.

Brands pay attention to supplying capacity when reviewing suppliers to initiate business, and guarantee that the suppliers are well informed to plan their production flows accordingly. Some brands such as Nordstrom arrange factory visits to check the resources available within the company, i.e., equipment, staffing and capacity, before they start working with new suppliers. Gap, a leading apparel brand, follows the practice of training employees on a regular basis to have a cordial and fruitful association with suppliers.

Another Dutch company, Mud Jeans, is actively involved in leasing jeans, providing customers with the options of preserving, swapping or returning their used ones for recycling. Monitoring and surveillance of working conditions and environmental management becomes the prime concern of this mobile technology. Likewise, Levi has come up with an innovative method of reducing water consumption in its finishing process by the creation of a water recycling system that saves millions of litres. It can thus be inferred that major brand are progressively marching towards ethical concepts by being more environmentally and socially responsible.

Simon Miller, a sustainable hand bag brand has attempted several measures such as utilization of organic mills and ozone technology during production aiming to reduce environmental impact and ensuring conservation of natural resources.

Verloop, an accessory brand follows the concept of trash to treasure by utilizing fabric scraps and excess materials to develop utilitarian accessories like winterwear accessories, slippers, bags and home décor artifacts.

Outland Denim, Australian manufacturers of ethical, stylish high-quality jeans are leading by example by contributing to both environmental and social causes via Denim Project. As far as manufacturing process is concerned, the brand utilizes organic cotton, natural dyestuff and sustainable packaging. As part of social obligations, the brand via its denim project shoulders the responsibilities related to education, employment and empowerment of Cambodian community, and victims of trafficking. The project deems to equip and empower rehabilitated girls rescued from industries offering them means of self-reliance and livelihood.

8 Capsule/Collaborative Collection—A Step Towards Ethical Work Practises

The buying behaviour and preferences of consumers has changed in recent times with consumers refraining from mindless shopping and becoming more conscious of their environmental responsibilities. Accordingly, today's aware consumer likes to be updated about the constituents of their clothing and are gradually drifting to eco-friendly clothing options. Accordingly, most of the textile manufacturing companies and major fashion brands have switched over to sustainable principles in their supply chain considering the inclination of consumers towards sustainability. The first path way for meeting sustainable goals is at the raw material stage and thus natural and synthetic fibres that demand wasteful utilization of natural resources are being replaced by sustainable alternatives like soya bean, pineapple, hemp, jute, bamboo, recycled polyester and organic cotton to name a few.

Major apparel and accessory brands sensing the consumers' preferences and the worldwide clamour for sustainable and ethical fashion, have been developing trans-seasonal, capsule collections encouraging the consumers to mix and match their ensembles.

The effluent apparel and accessory designers and brands are gradually stepping towards sustainable approaches and initiatives as part of their social and environmental responsibility. Accordingly, the fashion brands that were once ambassadors for fast fashion are investing in brand positioning and lucrative collaborations. H&M has been trailblazer in following the concept of fashion collaboration.

H& M in collaboration with Sabyasachi has launched a travel themed capsule collection that offers grandeur of bohemian era featuring warm colours, shimmering, golden embroidered fabrics and flowing silhouettes. The capsule collection flaunts of multicultural Indian brigade celebrating the prowess and affluence of Indian craftsmanship (Fig. 1).

Likewise, H&M has collaborated with designer Sandra Mansour for another capsule collection inspired from natural elements. The collection focuses on spreading hope and optimism in difficult times using natural elements like sunflower that symbolize cycle of life. The designers have garnered inspiration from poetry and painters while selecting trimmings and fabrics for the collection. Accordingly, embroidered organza, jacquards and dark laces rule the collection replacing the staples.

Moreover, the fashion brands in the pursuit to follow sustainable principles have been advocating the concept of local for vocal with designers like Anita Dongre, Rahul Mishra working in close association with artisans to incorporate the hand curated skills of artisans in their couture collections thereby promoting the rich legacy of traditional textiles (Figs. 2 and 3).

9 Gender Fluid Clothing—An Approach Towards Ethical Practises

The inception of gender fluid, gender neutral or androgenous fashion has met overwhelming response from Gen Z who are experimental, fervent and flamboyant and prefer shopping outside the confined gendered domains. Gender neutral clothing can

Fig. 1 Collaborative capsule collection

be referred to as self-fashioned and self-constructed clothing unrestricted by traditional concept of menswear or women wear binary. The coined term debunks the myth of social construct of gender-based clothing categorization and successfully disregards the association between pants and males and females and skirts thereby giving flexibility to wearer to choose attire devoid of any social stigma. In so doing, the gender neutrals support the cause like rejecting societal norms and providing autonomous expression to wearer to dress up without any inhibitions unlike the mainstream fashion. Oversized, formless, shapeless, form fitting, solid coloured baggy pants to oversized t shirts and sneakers, biker jackets are some classic examples of gender-neutral ensembles.

A gamut of values-driven brands like Older brother, Kirrin Finch, Big Bud Press, Riley Studio, Ijji, Organi Customs, Tomboy X, Lonely Kids Club, Official Rebrand, etc. in the pursuit to incorporate sustainability and minimalistic approach are offering gender fluid clothing that boasts of belonging to everyone's closet irrespective of gender (Fig. 4). Moreover, brands like H& M and Zara have also been experimenting with genderless collection with the launch of Denim United Line and Ungendered collection, respectively.

The brands endorsing the gender-neutral ensembles prefers keeping the attire natural with due consideration to adoption of sustainable approaches like cultivation of unique hibiscus-based dyes, sustainable wood bark by Older brother, utilization of pre-worn items for design and development of customized upcycled garments by

Fig. 2 Ethical fashion brands

Official Rebrand. The brands are wary of their social responsibilities as well and quest for narrating social awareness among masses such as mental health, woman empowerment, gender inclusivity. Lonely Kids Club committed to social awareness has been working on these social issues for creation of safe space for both genders. Likewise, Tomboy X, a unisex undergarments manufacturer is another gender-inclusive brand that refrains the usage of harmful chemicals and pioneered in water recycling program for water-based methods. Origami Customs, another gender-neutral brands collaborates with closed loop and family-owned partners to manufacture handmade swimwear and lingerie that are ethically produced utilizing deadstock, indigenous and recycled fabrics. Ijji, a genderless California based garment brand primarily procure natural and sustainable fibres such as wool, organic cotton, etc. for their exclusive design collections. Riley Studio another eco-conscious apparel brand utilizes recycled and regenerative materials like recycled polyester, ECONYL yarn and Recover Yarn to develop a range of casual and semi formal attires. Furthermore, the brand is transparent in sharing factory and supplier details associated with their supply chain. Big bud Press prioritizes local sourcing and ethical manufacturing for a range of unisex apparels, bags and accessories. Kirrin Finch known for their dapper style blazers and suits adhere to ethical practises by utilization of organic dyestuff and sustainable materials.

Fig. 3 Sustainable design collection

10 Convertible/Transformable Fashion

Convertible/Transformable fashion entails apparels and accessories capable of portraying multiple aesthetic and functional aspects (Fig. 5) via varied manipulation techniques like inclusion of concealed fasteners, detachable components, experimenting with design elements like addition of godets, eyelets, draping, binding, gathering, shirring, etc. Transformable garments in contrast to traditional clothing showcase at best two or more different appearances sharing certain features and functionalities identical to virgin attire thereby refraining the consumers from mindless, unnecessary purchases owing to multi-functional and aesthetic appeal offered by the ensemble. The product usage frequency and extension of the products' life cycle can be accomplished by transformability with the same attire being transformed to present an altogether different style or look by minimal manipulation, consequently minimizing the textile waste and decreased production volumes. The transformable garments in order to hit mainstream market like traditional attires should fulfil the functional, hedonic and social expectations of consumers. Functional attributes of clothing encompass ease of matching and layering, comfort, utility, easy care properties, serviceability and durability. Hedonic aspects are related to experimentation with varied clothing styles while modesty and suitability of clothing as per consumers' preferences, lifestyle and demographics constitute the social aspects of clothing.

Fig. 4 Gender neutral clothing

The concept of transformable/convertible fashion has served best of both worlds for consumers and retailers alike. The fashion connoisseurs who wish to embrace new styles but are budget conscious as well are served with multiple options in the same transformable attire. The product competitiveness and exclusivity in marketplace increases with incorporation of transformable concepts in garments thus enabling the retailers to strategize, outshine and make profits by satisfying consumers' aspirations for novelty and versatility.

A variety of brands namely 180DEGREES, JOLiER, Ximena Valero Corporation, WORKHALL studio, Y-dress design, Hip KnoTies follow the core values of transformability, reversibility, size adjustability and sustainability in their design collections offering innovative multi-uses to the consumers (Fig. 5).

Fig. 5 Convertible apparels and accessories

11 Conclusions

The fashion industry has been adversely affected by the recent economic upheaval and slow down owing to pandemic breakdown. The focus of millennials, Gen Z and masses of varying socio-economic categories has dwindled from vanity and fancy clothing to functional and comfortable ensembles. Accordingly, ethical fashion along with inception of innovative concepts like capsule and gender fluid clothing are at forefront with fast fashion archaic in present times. Ethical fashion pertains to ethical protocol followed in the entire supply chain to ensure no detrimental impact on environment and social well-being of stake holders and those associated with supply chain.

The entire textile and garment supply chain can be transformed by incorporation of sustainable principles thus eliminating the wastage in the entire production process from cradle to grave stage of the products. Furthermore, major apparel and accessory brands sensing the consumers' preferences and the worldwide clamour for sustainable and ethical fashion, have been adopting sustainable approaches in order to adapt to changing scenario. A gamut of values-driven brands are keen in adoption of ethical practises by following minimalistic approaches like capsule collection that offers the essence of limited, exclusive version of designer's collection or transforming the conventional gender-based clothing into genderless or gender neutral ensembles that boasts of belonging to everyone's closet irrespective of gender. Likewise, the ethical and sustainable principles can be integrated in textile and fashion supply chain refraining consumers from mindless shopping by offering multiple functionalities and aesthetics in the transformable or convertible garments.

References

1. https://www.masterclass.com/articles/capsule-collection-explained#what-is-the-difference-between-a-capsule-collection-and-a-capsule-wardrobe
2. https://pixelphant.com/blog/trending-fashion-trends-this-year
3. https://about.puma.com/en/newsroom/brand-and-product-news/2021/06-14-2021-
4. https://www.indianretailer.com/news/indian-terrain-leads-sustainability-in-high-street-fashion-with-fairtrade-capsule.n10242/
5. https://www.vogue.in/fashion/content/1111-eleven-eleven-has-a-gender-neutral-capsule-collection-on-matchesfashion
6. https://in.fashionnetwork.com/news/puma-unveils-ferrari-premium-capsule-collaboration,649857.html
7. https://fashionretail.blog/2017/09/12/collaborations-in-fashion-capsule-collection/
8. https://www.licenseindia.com/archives/article/how-retailers-are-reprising-licensing-with-capsule-collection
9. https://www.nishmagazine.com/2020/08/11/hm-capsule-collection-2020/
10. https://fashionretail.blog/2017/09/12/collaborations-in-fashion-capsule-collection/
11. https://consciousfashion.co/guides/eco-capsule-wardrobe-brands
12. Cline EL (2009) The Conscious Closet
13. Hausmann K (2015) Fashion capsule wardrobe essential: stylish work
14. Loderaback SE (2015) Secrets of the Capsule Wardrobe: how to find your personal style & create a happy, confident closet
15. Jhanji Y (2020) Sustainable fashion material procurement, Supply chain management and logistics in the global fashion sector: the Sustainability Challenge, Nayak R (ed). Woodhead Publishing, Cambridge
16. Jhanji Y (2021) Problems and hazards of textile and fashion waste and measures for effective waste management. In: Nayak R (ed) Waste Management in Textile & Fashion Industry. Elsevier
17. https://remake.world/stories/news/ethical-fashion-vs-sustainable-fashion-whats-the-difference-how-do-they-intersect/
18. https://ethicalmadeeasy.com/what-is-ethical-fashion/
19. https://www.thegoodtrade.com/features/what-is-ethical-fashion
20. Niinimäki K (2015) Ethical foundations in sustainable fashion. Text Cloth Sustain 1:3. https://doi.org/10.1186/s40689-015-0002-1
21. https://99designs.com/blog/tips/ethical-design/
22. https://www.sustainablejungle.com/sustainable-living/ethical-sustainable-fashion/
23. https://www.supplychainbrain.com/blogs/1-think-tank/post/30725-six-steps-to-an-ethical-and-sustainable-supply-chain
24. https://www.fibre2fashion.com/industry-article/8592/simple-ways-to-commit-to-sustainable-fashion-and-avoid-the-fast-fashion-habit
25. Varghese AM, Mittal V (2018) Biodegradable and Biocompatible Polymer Composites
26. https://www.fibre2fashion.com/industry-article/8701/hemp-fibre-for-high-quality-textile
27. https://www.genpact.com/insight/blog/responsible-sourcing-where-ethics-meet-growth
28. Lang C, Wei B (2019) Convert one outfit to more looks: factors influencing young female college consumers' intention to purchase transformable apparel. Fash Text 6:26. https://doi.org/10.1186/s40691-019-0182-4

Printed in the United States
by Baker & Taylor Publisher Services